Silk reeling and testing manual

FAO
AGRICULTURAL
SERVICES
BULLETIN
136

by
Yong-woo Lee
National Sericulture and Entomology Institute
Seoul
Republic of Korea

2005
Published by Arrangement with the
Food and Agriculture Organization
of the United Nations
by
DAYA PUBLISHING HOUSE
Delhi - 110 035
for Sale and Distribution only in South-East Asia

Food
and
Agriculture
Organization
of
the
United
Nations

Rome, 1999

"This book was originally published by the Food and Agriculture Organization of the United Nations (FAO) as FAO Agricultural Services Bulletin No. 136."

"The designations employed and the presentation of material in this publication do not imply the expression of any opinion whatsoever on the part of the Food and Agriculture Organization of the United Nations concerning the legal status of any country, territory, city or area of its authorities, or concerning the delimitation of its frontiers or boundaries."

"The designations "developed" and "developing" economies are intended for statistical convenience and do not necessarily express a judgement about the stage reached by a particular country, territory or area in the development process."

First Indian Edition 2005

ISBN 81-7035-398-X

Published by : **Daya Publishing House**
1123/74, Deva Ram Park
Tri Nagar, Delhi - 110 035
Phone: 27383999
Fax: (011) 23244987
e-mail : dayabooks@vsnl.com
website : www.dayabooks.com

Showroom : 4762-63/23, Ansari Road, Darya Ganj,
New Delhi - 110 002
Phone: 23245578, 23244987

Printed at : **Chawla Offset Printers**
Delhi - 110 052

PRINTED IN INDIA

FOREWORD

The silk reeling industry in general, covers a wide range of techniques. These include handling of fresh cocoons, cocoon drying and storage, cocoon boiling and reeling, raw silk re-reeling and finishing, management of water quantity and quality, utilization of by-products and maintenance of machinery. Both continuous improvement and effective dissemination of technology for silk reeling and testing are essential to meet the increasing global demand for quality raw silk and fine silk fabric, as ever more developing countries become interested in both cocoon and silk production. The author has tabulated fundamental data and explained the basic principles of silk reeling technology so that they can be easily understood by those do not have any previous experience or knowledge of silk reeling.

This manual can also be utilized as a reference guide for silk reeling managers and staff. Furthermore, the techniques described in this manual respond favourably to bivoltine cocoons in the temperate zones. Many tropical countries are trying to adopt bivoltine cocoon reeling, which produces more quantity and better quality silk than multivoltine cocoons.

This manual also provides information on what methods of silk reeling are to be used to produce good raw silk and silks fabric. It also contains the answer on how to adjust filature processes based upon the properties of raw material cocoons.

Morton Satin
Chief
Agro-industries and Post-harvest Management Service
FAO Agricultural Support Systems Division

TABLE OF CONTENTS

PAGE

LIST OF FIGURES

LIST OF TABLES

PAGE

PAGE

CHAPTER 1

INTRODUCTION

The origins of sericulture and silk production are closely associated with the emergence of China as one of the great civilizations. It is believed that sericulture evolved gradually and by the middle of the third millennium BC was already being used by humanity.

Silk is an animal fibre, produced by caterpillars belonging to the genus *Bombyx*. A single silk filament is the product of a series of stages derived from the cultivation of mulberry trees for feed to the propagation of the domesticated silkworm, *Bombyx mori*. During the caterpillar phase, the worm wraps itself in a liquid protein secreted by two large glands in its head. This secreted protein hardens upon exposure to the air. The resulting filament is bonded by second secretion, sericin, which forms a solid sheath or cocoon. Under natural conditions, a moth eventually breaks through the cocoon. In sericulture, the larva is killed in the cocoon by steam or hot air in the chrysalis stage before its metamorphosis. Sustained heat processing softens the hardened sericin so that the filament can be unwound.

The silk filament is a continuous thread of great strength measuring from 500-1 500 metres in length. Single filaments are too thin for utilization. For production purposes, several filaments are combined with a slight twist into one strand. This process is known as "silk reeling or filature".

Silk is a premium priced agricultural commodity, although its sheer volume is less than 1 percent of the market for natural textile fibres (Table 1). The international demand for high quality silk has multiplied. Appropriate cocoon-drying techniques and reeling operations are vital to supply good quality silk. Although new technologies are applied among major silk producing countries in Asia, many newcomers to the industry including non-traditional producers have not been transitioned to the methods that ensure high quality silk for export.

World production of raw silk reached nearly one hundred thousand tonnes in 1993. In contrast to strong declines in Japan and the Republic of Korea to roughly 3 900 and 500 tonnes respectively. China for the first time surpassed the 66 000 tonnes and India raised its output to 14 000 tonnes. Outside Asia, Brazilian output exceeded 2 500 tonnes per year, with the then USSR output lingering around 3 000 tonnes yearly. When compared to these three countries, production figures in some traditional sericulture countries have declined.

Table 1 – World production of textile fibres (in >000 tonnes)

Category/year	1975	1980	1985	1990	1995
Cotton	11 809	13 981	17 540	18 447	18 200
Wool	1 502	1 666	1 673	1 897	1590
Cellulose fibres	2 959	3 242	3 000	2 988	2 375
Synthetics	7 346	10 476	12 515	15 830	17 795
Silk	55	68	68	85	91

Recently, China experienced a striking decline in output to 59 000 tonnes in 1996; however a slight recovery was anticipated in 1997. At this time, China's production will be sustained around 60 000 tonnes per year. Output in India also diminished from nearly 14 000 tonnes in 1994 to less than 13 000 tonnes by 1996 (Table 2). Still over 70 percent of global output are attributed to China, followed by India as the second largest producer.

Table 2. Production of raw silk in the world (Unit : M/T)

Year / Country	1938	1970	1980	1985	1990	1992	1993	1994	1995	1996
Brazil	33	259	1 170	1,554	1 692	2 298	2 328	2 538	2 466	2 242
Bulgaria	180	180	252	180	180	180	180	180	180	180
China	4 853	11 124	23 460	33 000	40 800	51 000	60 540	66 060	64 584	59 000
D.P.R. Korea	-	-	690	600	1 380	1 380	1 380	1 380	1 104	1 104*
Greece	255	40	36		8	8	8	8	8	8
India	691	2 258	3 960	6 960	11 484	13 002	13 434	13 914	12 882	12 600*
Indonesia	-	-	31	17	17	17	17	17	17	17
Iran	210	210	432	432	432	432	432	432	432	432
Italy	2 738	310	10	12	12	12	12	12	12	12
Japan	43 152	20 515	16 152	9 594	5 718	5 085	4 254	3 900	3 180	2 579
Lebanon	21	21	10	10	10	10	10	10	10	10
Rep. of Korea	1 824	3 024	2 278	1 625	948	898	666	492	342	100
Rumania	15	15	126	126	126	126	126	126	126	126
Thailand	-	-	300	828	1 470	1 518	1 650	1 788	1 068	1 068
Turkey	213	120	258	312	210	210	210	210	210	210
ex-USSR	1 900	1 900	4 410	3 738	4 094	4 092	3 738	2 850	2 952	2 900*
Vietnam	178	178	72	72	498	600	900	1 500	1 800	1 000*
Others	237	844	668	164	41	82	97	81	103	82
TOTAL	56 500	41 000	55 315	59 232	69 120	80 934	89 982	95 498	91 476	83 670

Source: I.S.A. Bulletin (* Estimated)

CHAPTER 2

CHARACTERISTICS OF THE COCOON

2.1 Physical characteristics of cocoons

The silk glands of the *Bombyx mori* are structured like tubes consisting of a Posterior, Middle and Anterior section. The Posterior is long and thin. The Middle is short with a diameter measuring 3-4 mm. The Anterior is extremely thin, leading to the spinneret in the head of the larvae (See Appendix, Figure 1) from which the silk is excreted.

Fibroin is secreted in the Posterior and transferred by peristalsis to the Middle section, which acts as a reservoir. Here it is stored as a viscous aqueous solution until required for spinning. The majority of the sericin is created within the walls of the Middle section. In fact, these two proteins are reserved side by side in the Middle section without mixing one into the other. The fibroin core is covered with a layer of sericin and the secretions from the two proteins join at the junctions where the sericin is fused into one layer. The Filipis glands discharge a liquid protein. To form its cocoon, the silkworm draws out the thread of liquid protein and internally adds layer after layer to complete this protective covering.

Colour

Colour is a characteristic particular to the species. It is the presence of pigments in the sericin layers, which cause the colour. This colour is not permanent and washes away with the sericin during the degumming process (see Chapter 5). There are diverse hues of colour including but limited to white, yellow, yellowish green and golden yellow.

Shape

Cocoon shape, as colour, is peculiar to the given species. At the same time shape can be affected by the execution of the mounting process (see Chapter 3), especially during the cocoon spinning stage. Generally, the Japanese species is peanut-shaped, the Chinese elliptical, European a longer elliptical and the polyvoltine species spindle-like in appearance. Hybrid cocoons assume a shape midway between the parents for example, the case of a longer ellipsoid or shallowly enclosed peanut form (see Appendix, Figure 3). The shape of cocoons assists in identifying the variety of species plus evaluating reelability (see Reelability, page 6).

Wrinkle

The deflossed cocoon has many wrinkles on its surface. Wrinkles are coarser on the outer layer than within the interior layer. The outline of the wrinkle is not uniform, but various according to species and breeding conditions. Spinning employs high temperature and low humidity settings, which render fine wrinkles or cotton-like textures of, cocoon layers. These provisions discourage the agglutination of the baves resulting from accelerated drying. It is recognized that coarse wrinkled cocoons reel poorly.

Size

Cocoon size or volume is a critical characteristic when evaluating raw materials. The size of the cocoon differs according to silkworm variety, rearing season and harvesting conditions. The number of cocoons per litre, ranging between 60 and 100 in bivoltine species calculates size. Multivoltine species measure considerably higher.

Cocoon weight

The most significant commercial feature of cocoons is weight. Cocoons are sold in the marketplace based on weight as this index signals the approximate quantity of raw silk that can be reeled. The whole weight of a single cocoon is influenced by silkworm species, rearing season and harvest conditions. Pure breeds range from 2.2 to 1.5 g, while hybrid breeds weight from 2.5 to 1.8 g. In nature, the weight of a fresh cocoon does not remain constant but instead continues diminishing until the pupae transforms into a mother and emerges from the cocoon. This weight occurs gradually as moisture evaporates from the body of the pupae and as fat is consumed during the metamorphosis process (Table 3).

Table 3. Daily loss in weight of fresh cocoons

Days after mounting	6	7	8	9	10	11	12	13
Days after pupation	2	3	4	5	6	7	8	9
Index of fresh cocoon weight	100	99.4	98.8	98.3	97.7	97.0	96.1	95.1

Thickness/weight of cocoon shell

The thickness of the cocoon shell is not constant and changes according to its three sections. The central constricted part of the cocoon is the thickest segment, while the dimensions of the expanded portions of the head are 80 to 90 percent of the central constricted (Table 4).

The weight of the silk shell is the most consequential factor as this measure forecasts raw silk yield. As with other characteristics introduced in this chapter, shell weight differs in correspondence to varieties of silkworms. Further, weight is also influenced by the type of technology used for rearing and mounting. In practice, uni and bivoltine species produce heavier shell weights than multivoltine species.

Hardness or compactness

Cocoon hardness correlates to shell texture and is affected by cocoon spinning conditions. For instance low humidity during the mounting period (see Chapter 3) makes the cocoon layer soft, while high humidity makes it hard. The degree of hardness also influences air and water permeability of cocoons during boiling. A hard shell typically reduces reelability (during the cocoon reeling process), while a soft-shell may multiply raw silk defects. In short, moderate humidity is preferred for good quality cocoons.

Table 4 Variations of shell thickness in different parts

Varieties		A		B		C	
Parts		micron	index	micron	index	micron	index
Head	extreme end	336	47	302	43	228	40
	expanded	594	83	596	85	480	84
Central constricted		712	100	700	100	572	100
Tail	expanded	568	80	590	84	466	81
	extreme end	382	54	333	48	348	43

Shell percentage

As the entire cocoon including the pupa is sold as part of the raw material, it is essential to quantify the ratio of the weight of the silk shell versus the weight of the cocoon. This is calculated in the formula:

$$\frac{\text{Weight of the cocoon shell}}{\text{Weight of the whole cocoon}} \times 100$$

This value gives a satisfactory indication of the amount of raw silk that can be reeled from a given quantity of fresh cocoons under transaction. The calculation assists in estimating the raw silk yield of the cocoon and in deriving an appropriate price for the cocoons. The percentage will change based on the breed of the silkworms, rearing and mounting conditions. Percentage rates are altered based on the age of the cocoons (see cocoon weight) as the pupa loses weight as metamorphosis continues. In newly evolved hybrids, recorded percentages are 19 to 25 percent, where male cocoons are higher than female cocoons.

Raw silk percentage

This index is the most important for the value of the cocoon as it has a direct impact on both the market price of cocoons and the production costs of raw silk. The normal range is 65 to 84 percent for the weight of the cocoon shell and 12 to 20 percent for the weight of the whole fresh cocoon.

normal range is 65 to 84 percent for the weight of the cocoon shell and 12 to 20 percent for the weight of the whole fresh cocoon.

Filament length

Equally important as the percentage of silk shell is measuring the length of the bave contained in the shell. The factor determines the workload, rate of production, evenness of the silk thread and the dynamometric properties of the output. The length of cocoon filament corresponds to the varieties of silkworms. Range of total length is from 600 to 1 500 m of which 80 percent is reelable while the remainder is removed as waste.

Reelability

Reelability is defined as the fitness of cocoons for economically feasible reeling. Industry practice measured the case with which the cocoon yields the bave in reeling. Poor reelability causes a variety of production problems such as halts in production due to filament breakage and high degrees of waste product. Reelability is greatly affected by careful action during cocoon spinning, drying, storage, pre-processing, reeling machine efficiency and operator skill.

$$\text{Reelability (\%)} = \frac{\text{Number of reeled cocoons}}{\text{number of ends feeding}} \times 100$$

Recent statistics show an average reelability of percent for good cocoon varieties. The measured range is from 40 to 80 percent with serious deviations depending on the type of cocoon. Note that stained cocoons generally have poor reelability.

Size of cocoon filament

The measure denier expresses the size of silk thread. A denier is the weight of 450 m length of silk thread divided into 0.05 g units. The diameter of the bave is not constant throughout its length, instead changes according to its position in the bave shell. At the coarsest section of cocoon filament from 200 to 300 meters, the denier increases. Once more these dimensions become finer and finer as the process approaches the inside layer (see Figure 1). The average diameter of cocoon filament is 15 to 20 microns for the univoltine and bivoltine species.

Defects

A series of minor defects may be found in cocoon filament such as loops, split-ends, fuzziness, nibs and hairiness (Figure 2). While these defects are observed among silkworm varieties, mounting conditions seem to contribute to their incidence. These filament defects directly affect raw silk quality. It is not recommended that silk varieties graded below 90 percent in the Neatness Chart be used.

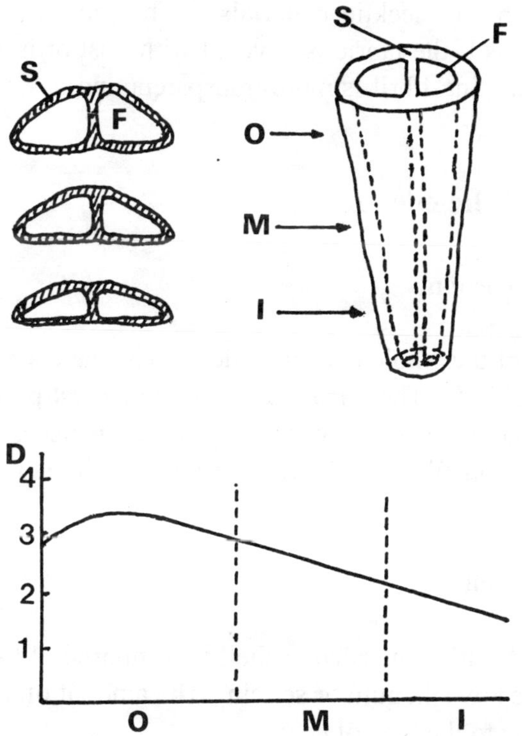

Figure 1 – Size curvature of cocoon bave
S: sericin; F: fibroin; O: outer layer; M: middle layer; I: inner layer; D: size (denier)

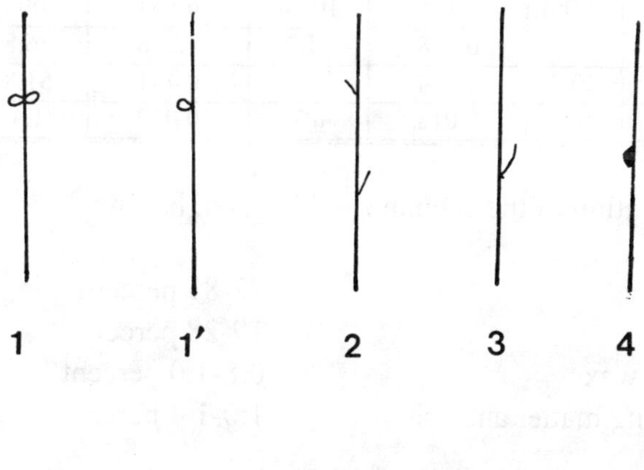

Figure 2 – Neatness defects
(1) (1') Loop (2) Hairiness (3) Split-ends (4) Nibs

Hair-like projections in the silk fibre are called Lousiness. Lousiness is more prevalent in baves produced by silkworms, which have been overfed in their fifth stage of rearing. Lousiness is found less in breeds of silkworms, which spin finer bave. Another factor promoting lousiness is mounting of over-mature larvae. This defect poses serious problems to silk fabric manufacturers, in particular those producers of smooth satin and necktie materials. When fabrics woven with these defects are dyed, it looks as if the fabric is covered with dust or is a paler shade than the rest. In fact, the protruding fibril is more transparent and has a lesser capacity to absorb dyes.

2.2 Composition of the cocoon

Composition of a whole cocoon

The composition of the whole cocoon is defined as the cocoon shell, pupa and cast off skin shown in Table 5. The pupa makes up the largest portion of its weight. Note that much of the cocoon content is water; therefore it is necessary to remove the water to improve the cocoon filament for reeling and to better preserve the cocoon over a long period.

Composition of cocoon shell

The silk filament forming the cocoon shell is composed of two brins (proteins) named fibroin and covered by silk gum or sericin. The amount of sericin ranges from 19 to 28 percent according to the type of cocoon.

Table 5. Composition of the cocoon

Weight	Fresh Cocoon				Dried Cocoon			
	Race A		Race B		Race A		Race B	
	Actual number (g)	Ratio (%)	Actual Number (g)	Ratio (%)	Actual Number (g)	Ratio (%)	Actual Number (g)	Ratio (%)
Cocoon	2.181	100.0	2.156	100.0	0.851	100.0	0.888	100.0
Cocoon shell	0.404	18.5	0.458	21.2	0.398	46.8	0.452	50.0
Pupa	1.765	80.9	1.684	78.1	0.441	51.8	0.422	47.5
Cast-off skin	0.012	0:6	0.014	0.7	0.012	1.4	0.014	1.6

The composition of the cocoon shell is given below:

Fibroin	72-81 percent
Sericin	19-28 percent
Fat and wax	0.8-1.0 percent
Colouring matter and ash	1.0-1.4 percent

Usually the sericin content of the cocoon shell is at the maximum level at the outside layer 1 becoming progressively lower at the middle layers 2 and 3 and the absolute minimum at the inside layer 4 (Table 6).

Table 6. Sericin content to different layers of cocoon shell

Cocoon layers	Race A (%)	Race B (%)	Race C (%)	Race D (%)
Outside 1	31.40	32.08	34.13	33.15
↑ 2	23.45	29.29	27.50	27.71
↓ 3	20.11	22.22	23.96	23.47
Inside 4	18.12	20.63	21.54	21.33

2.3 Properties of silk

Structural features of silk

The silk of *Bombyx mori* is composed of the proteins fibroin and sericin, matter such as fats, wax, sand pigments plus minerals.

Fibroin in the *Bombyx mori* comprises a high content of the amino acids glycine and alanine, 42.8 g and 32.4 g respectively as shown in Table 7.

The key amino acids in sericin are serine (30.1 g), threonine (8.5 g), aspartic acid (16.8 g) and glutamic acid (10.1 g) (see Table 7).

Table 7. Amino acid composition of Fibroin and Sericin (Kirimura, 1972)

Amino acids	Fibroin	Sericin	Amino acids	Fibroin	Sericin
Glycine	42.8	8.8	Glutamic acid	1.7	10.1
Alanine	32.4	4.0	Serine	14.7	30.1
Leucine	0.7	0.9	Threonine	1.2	8.5
Isoleucine	0.9	0.6	Phenylalanine	1.2	0.6
Valine	3.0	3.1	Tyrosine	11.8	4.9
Arginine	0.9	4.2	Proline	0.6	0.5
Histidine	0.3	1.4	Methionine	0.2	0.1
Lysine	0.5	5.5	Tryptophan	0.5	0.5
Aspartic acid	1.9	16.8	Cystine	0.1	0.3

Values are given as gram of amino acid per 100 g of protein.

Sericin is a complex protein composed of three distinct components (I, II and III) of which sericin III is the interior layer directly adjacent to the fibroin core. The sericin I outer lay is the most soluble of the three constituents, while sericin III is difficult to dissolve. Viewed as a cross section, the brins have the appearance of equilateral triangles with rounded corners that face each other at their respective bases (Figure 3).

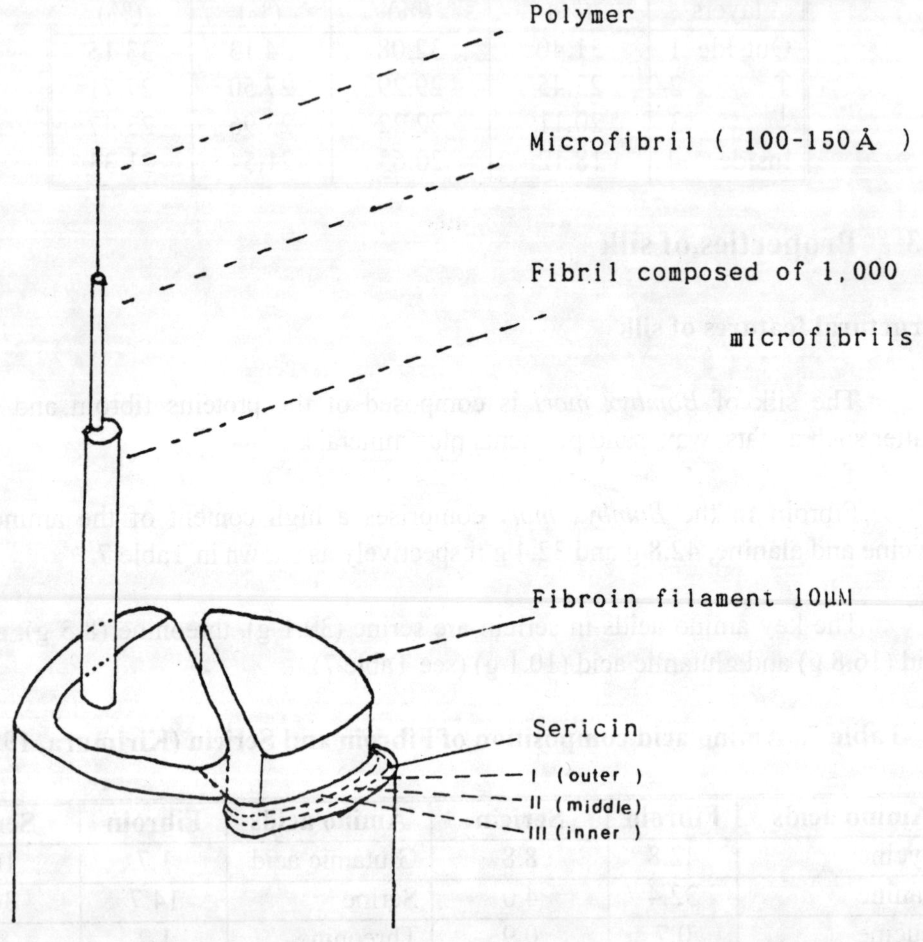

Figure 3. Texture of the silk thread

When the brin is crushed, it splinters into numerous minute fibrils revealing the actual structure of the brins. The thickness of each fibril is less than one micron and they are parallel to the axis of the fibre. A single fibril contains many microfibrils which, when examined with an electron microscope, have a diameter of approximately 100 Å per microfibril. Microfibrils contain micelles, which are separated into crystalline and amorphous segments.

Physical and chemical properties of silk

1. Specific gravity

The bave specific gravity on average of sericin and fibroin measures from 1.32 to 1.40. Generally, the specific gravity of sericin is slightly higher than that of fibroin (See Raw silk, Table 8).

Table 8. Specific gravity and tensil strength of various fibres

Fibres	Specific gravity	Tenacity (g/denier)	Elongation (%)
Raw silk	1.32-1.40	2.6-4.8	18-23
Degummed silk	1.30-1.38	-	-
Wool	1.30-1.40	1.2-1.5	30-48
Cotton	1.52-1.60	3.2-4.8	7-11
Flax	1.50-1.58	4.8-6.0	2-4
Nylon	1.14-1.17	4.5-5.0	25-30

2. Tenacity and elongation

Tenacity indicates the quantity of weight a given fibre can support before breaking. the typical tenacity of a bave is 3.6 to 4.8 g per denier (see Raw silk, Table 8). Degummed silk has greater tenacity than raw silk. Elongation defines the length to which a fibre may be stretched before breaking. Raw silk has an elongation of 18 to 23 percent of its original length. Excess moisture increases the elongation of silk, but decreases its tenacity.

3. Hygroscopic nature

$$\text{Denier (D)} = \frac{\text{Fibre weight (g)}}{\text{Fibre length (m)}} \times 9\,000$$

Moisture content and humidity are of critical importance to commercial silk production. Figure 4 illustrates the pattern of moisture regain where a hysteresis exists between the adsorption and desorption curves. Desorption measures a greater regain at a given relative humidity. For instance, given 65 percent RH, the adsorption regain value is 10 percent and the associated desorption value is 11.1 percent. Currently, 11 percent is the accepted moisture regain coefficient for silk; the mercantile weight of silk is derived based on this factor.

4. **Effect of light**

Continuous exposure to light weakens silk faster than cotton or wool. Raw silk is more resistant to light than degummed silk. It is advised that silk drapery and upholstery fabrics be protected from direct exposure to the light.

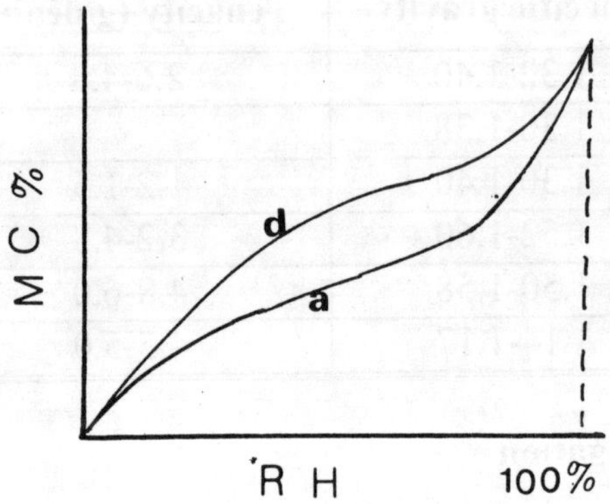

Figure 4. Hysteresis phenomenon
a : adsorption, d : desorption, R.H. : relative humidity, MC : moisture content

5. **Electrical properties**

Silk is a poor conductor of electricity and accumulates a static charge from friction. This trait can render it difficult to handle in the manufacturing process. This static charge can be dissipated by high humidity or by maintaining a R.H. of 65 percent at 25°C. Based on its insulating properties, silk is used extensively for covering wire in electrical equipment.

6. **Action of water**

Silk is a highly absorbent fibre, which readily becomes impregnated with water. Water, however, does not permanently affect silk fibre. Silk strength decreases about 20 percent when wet and regains its original strength after drying. The fibre expands but does not dissolve when steeped in warm water. Note that the fibre will also absorb dissolved substances present in water. This is the reason that special attention is given to the quality of the water utilized for reeling, washing, dyeing or finishing.

7. Effect of heat

If white silk is heated in an oven at 110°C for 15 minutes, it begins to turn yellow. At 170°C, silk disintegrates and at its burning points releases an empyreumatic odour.

8. Degradation by acids, alkalis

Treatment of silk fibres with acid or alkaline substances causes hydrolysis of the peptide linkages. The degree of hydrolysis is based on the pH factor, which is at minimum between 4 and 8. Degradation of the fibre is exhibited by loss of tensile strength or change in the viscosity of the solution.

Hydrolysis by acid is more extensive than alkali, and it has been postulated that acid hydrolysis occurs at linkages widely distributed along the protein chain, whereas in the early stages of the alkaline treatment, hydrolysis happens at the end of the chain. Hydrochloric acid readily dissolves fibroin especially when heated – and this is used mainly in studies of hydrolysis. Hot concentrated sulphuric acid, while rapidly dissolving and hyrolyzing fibroin, also causes sulphation tyrosine.

Nitric acid readily decomposes fibroin, due to its powerful oxidizing properties and concurrently causes nitration of the benzene nuclei. Organic acids have few effects at room temperature when diluted, but in a concentrated form fibroin may be dissolved, along with a certain amount of decomposition.

9. Proteolytic enzymes

Proteolytic enzymes do not readily attack fibroin in fibrous form apparently because the protein chains in silk are densely packed without bulky side chains. Serious degradation may be caused by water or steam at 100°C.

10. Oxidation

Reports regarding the oxidation of proteins are rather meagre since the reactions are very complex. Oxidizing agents may attack proteins in three possible points:

a) at the side chains,
b) at the N-terminal residues, and
c) at the peptide bonds of adjacent amino groups.

Hydrogen peroxide is absorbed by silk and is thought to form complexes with amino acid groups and peptide bonds. It has been demonstrated that hydrogen peroxide diminishes the tyrosine content and further that the peptide bonds are broken at the tryosine residues. Peracetic acid causes more rapid scission and produces more acid groups than peroxide.

11. Other agents

Chlorine attacks fibroin more vigorously than does sodium hypochlorite. The oxidation is mainly at the tyrosine residues.

CHAPTER 3

COCOON QUALITY AND CLASSIFICATION

3.1 Cocoon quality

A series of natural circumstances will produce variations in cocoon quality. Some of the most noteworthy include:

- differences in cocoon quality in the same batch;
- differences in cocoons produced in the same location by different farmers who have reared the same species;
- seasonal influences. In Japan for example, cocoons produced in the spring and late autumn are higher in quality than those in early autumn and summer;
- environmental conditions affect cocoon reelability such as temperature and humidity;
- processing technique in reeling will impact reeling efficiency as well as raw silk quality;
- bivoltine cocoons are superior quality compared to multivoltine silkworm species traditional farmed in tropical zones.

Recent silkworm cultivation now develops cross-breeds of multivoltine with bivoltine silkworms as a strategy to improve overall cocoon quality.

3.2 Factors influencing cocoon quality

This section presents the measures to be taken during silkworm rearing and mounting to obtain a better quality of cocoons with higher silk content, longer filament, better reelability and lower percentage of defective cocoons.

Temperature and humidity during mounting

Maintain temperatures at or near 25°C and relative humidity around 65 percent for silkworms to spin good quality cocoons with a high reelability.

Mounting device

Although different mount practices are employed among producer countries, rotary mounting frames provide good ventilation. The result is improved reelability of cocoons.

Harvesting and handling of fresh cocoons

Cocoons should be harvested only following complete pupation. In practice, the appropriate harvesting day would be the fifth day in tropical countries, and the seventh or eighth day in temperate countries, from the mounting date. If premature harvesting takes place, the silkworm will still be in its larval stage, weigh more, have fragile skin, and could likely be crushed, which would cause stains to the cocoon during handling and transportation.

Transport of fresh cocoons

After proper harvesting and removal of diseased or damaged cocoons, the fresh cocoons are taken to the market. For short distances, the farmer carries the cocoons in bamboo baskets or jut bags on his head or by bicycle. If the distance is longer, cocoons are transported in a van or a bus. Caution should be exercised when loading fresh cocoons on to the van to ensure that containers are loosely packed in tiers to avoid damage. Vibration and shock during long trips can spoil fresh cocoons. Cocoon quality is affected by steam produced while being transferred in a bag or basket. If there are defective cocoons (see section 3.3) fresh cocoon quality will be harmed.

Table 9 describes the impact of transportation on fresh cocoons. While it is advisable to avoid carrying cocoons over long distances, there are steps, which preserve silk reelability. First use of P.V.C. containers with 15 kgs capacity is recommended. Shock absorbers, such as sponge can prevent damage over long distances. To minimize the risk of heat deterioration, shipping should take place only during the night or early morning. Ideally, the fresh cocoons should arrive at the stifling unit within two to three days after harvest.

Table 9. Effect of transportation containers of fresh cocoons on reeling results (Song and Kim, 1974)

Transportation containers	Raw silk yield (%)	Cocoon reelability (%)	Reeling troubles per 10,000m (times)	Neatness (%)	Cleanness (%)
Bamboo basket with cotton bag	18.50	71.1	2.17	93.0	95.0
P.V.C. container with sponge (shock absorber)	18.75	75.0	1.83	96.5	96.5

3.3. Classification of cocoons.

When cocoons are sold at the market, price is assessed on the basis of cocoon quality. This is judged by grading shell percent, filament length, reelability and the percentage of defective cocoons. If the percentage of defective cocoons is high, the price will be affected. The next section outlines the characteristics of defective cocoons.

Defective Cocoons

1. Double cocoons

A double cocoon is spun by two worms, producing a filament, which does not unwind smoothly and tangles easily. As these cannot be reeled along with normal cocoons, double cocoons are used for manufacture of a coarse, non-uniform, stubby yarn called "doupion". Double cocoons may be caused by crowded mounting conditions, high temperatures, high humidity and mutation of silk species.

2. Inside stained cocoons (dead cocoons)

Dead cocoons are also known as melted cocoons. In this case, the pupa is dead and sticks to the inside shell of the cocoon causing a stain. Melted cocoons are called mutes because they do not make a sound when shaken. These cocoons are difficult to process and will result in silk, which is dull in colour.

3. Outside stained cocoons

These are recognized by a rusty colour spot on the cocoon shell caused by absorption of intestinal fluid/urine of the mature worm formed during mounting. Reelability is very poor in this case.

4. Printed cocoons

This defect may happen due to improper mounting frames; these are also called scaffold pressed cocoons.

5. Malformed cocoons

These are abnormally shaped cocoons, which may arise from species variation. This defect may be due to racial characteristics and breeding with mulberry leaves stained with agrochemicals.

6. Flimsy cocoons

Here, the shell is loosely spun in layers and has a low silk content. These cocoons are easily overcooked and produce waste.

7. Thin-end cocoons

One or both ends of the cocoon are very thin and risk bursting when processed. The cause of this defect may be attributed to species characteristics or improper temperature and humidity during rearing and mounting.

8. **Pierced cocoons**

This happens when a moth has emerged, been eaten by beetles or in the case of the emergence of a parasite. Pierced cocoons are unfit for reeling and can be used only for hand spinning or as raw material of machine spun silk yarn.

3.4 Cocoon testing and grading

In sericulturally advanced countries, cocoons are subjected to systematic testing and grading before sale. Prices are based on the quality of the cocoons. But in developing countries, there is no system for cocoon testing. Cocoons are sold on visual inspection and personal experience is relied upon in marketing of cocoons. No laws exist or compulsory testing and trading of cocoons. The result is that cocoons are simply auctioned or in certain instances, even sold at a price fixed by the Regional Departments of Sericulture. There is no direct correlation between price and quality of cocoons.

Major sericultural countries through the International Sericultural Commission (ISC) have studied an International Cocoon Classification System. However, diversion of silkworm species, techniques of breeding, silk reeling and other factors lead to the production of non-uniform cocoons. These parameters have challenged the effort to unify equipment and testing methods (see Table 10).

Cocoon testing and grading

Cocoon testing and grading may be accomplished with a compact automatic reeling machine as well as a multi-ends reeling machine, which is typical equipment in major sericultural countries. Testing methods used by major sericulture countries are displayed in Table 10.

The quantity of fresh cocoons, which are taken out of a lot for testing purposes depends on the actual weight of the lot on offer.

The cocoons on offer are divided into three batches.
Batch weighing up to 1 000 kgs
Batch weighing up to 2 000 kgs
Batch weighing up to 4 000 kgs

The sample size of fresh cocoons taken out of each batch for testing is as follows:

1st batch - 2.0 kgs
2nd batch – 4.5 kgs
3rd batch – 6.0 kgs

In the case of dry cocoons, the quantities taken out for testing from each batch are as follows:

1st batch up to 400 kgs of dry weight – 0.8 kgs taken out for testing.
2nd batch up to 800 kgs of dry weight – 1.8 kgs taken out for testing.
3rd batch up to 1 600 and over - 2.4 kgs taken out for testing.

1. Drying of the cocoon test sample.

The cocoons received must be dried as soon as possible and to an acceptable degree. To do this, the moisture content of the fresh pupa and the cocoon shell percentage must be measured. The percentage of drying is calculated according to the formula.

Table 10. Cocoon classification systems of major sericultural countries

CHINA	INDIA	JAPAN	KOREA
Visual and mechanical test	Visual test	Mechanical test (Auto reeling)	Mechanical test (Auto and multi-ends reeling)
A. Visual inspection - cocoon shell weight (g/20 p.c.s. of bivoltine or 40 p.c.s. of polyvoltine) - 1 ~11 grade : 9.0-7.0 g (Jiangsu) 9.4-6.4 g (Sichuen) B. Mechanical test (Multi-ends reeling) - Length of non-broken cocoon filament (major item) : 1~20 grades (950-340 m) - Percentage of good cocoons (auxiliary item) : 1~ 8 grades (94-82%) - Size of cocoon filament and percent of inside stained cocoons (correction items)	A. Estimated Renditta constant = ---------------------------- shell % of cocoon lot · constants : 165-133 shell%= (weight of 100 shells / weight of 100 cocoons) x 100 B. Cocoon pricing cost of cocoon per kg. = Kakame cost / Renditta - Kakame cost = silk price + income from by-products - (cost of manufacture + profits)	A. Cocoon testing items - Raw silk percentage of cocoon (%) - Percentage if eliminated cocoon (%) - Reelability percentage (%) B. Classification of cocoons - Grade : Reelability percentage 5A : 100-85 4A : 84-80 3A : 79-75 2A : 74-70 A : 69-65 B : 64-60 C : 59-55 D : 54-50 E : 49-	A. Cocoon testing items - Raw silk percentage of cocoon (%) - Percentage of eliminated cocoon (%) - Length of cocoon filament (m) B. Grading of cocoon - Length of cocoon filament 10 class (33.5-42.5 point) - Cocoon reelability percent 10 class (43.5-52.5 point) Final grade (1+2) A : over 90 B : 88-89 C : 86-87 D : 84-85 E : below 83

Drying percent of cocoon = (0.0115 x mc of fresh pupa – 0.2104)
x (percent of cocoon shell 1.15)
x (mc of fresh cocoon + 115)
(mc : moisture content of the cocoons)

Drying should be accomplished in one continuous process where the temperature is gradually decreased from 98°C to 60°C until the required ratio is obtained.

2. **Test to calculate percentage of eliminated cocoons**

This evaluation usually happens on a table under natural light. If natural light is insufficient based on time of day or weather conditions, elimination takes place under artificial light of 500 lux. to calculate the percentage of eliminated cocoons according to terms and conditions defined by cocoon classification, the following types of cocoons must be removed: double cocoons, thin-end cocoons, scaffold marked cocoons, malformed cocoons, flimsy cocoons.

3. **Batching of cocoon sampled for reeling**

After elimination of the bad cocoons, the remaining cocoons are batched for the reeling test as follows:

Batch	Sample cocoon (fresh weight)	Cocoon for reeling (dried weight)	Preliminary cooking
	kg	grams	grams
1st	2.0	300 x 2	80
2nd	4.5	300 x 2	80 x 2
3rd	6.0	300 x 3	80 x 2

4. **Cooking of the cocoon sample**

A small sample of cocoons must be cooked to determine the correct cooking conditions for the specific batch. Once these parameters are established, actual cooking of the entire sample batch can be completed.

5. **Reeling of cocoon sample**

Reeling should be carried out under the following conditions:

Item	Multi-end reeling	Automatic reeling
Temperature of groping end part	85ºC	80ºC
Reeling velocity	90 m/min	160 m/min
No. of reeling silk ends (per basin)	10 pcs.	3 pcs.
Length of croissure	10 cm	8 cm
No. of cocoons or objective size per reeling thread	8 pcs.	21 denier

6. **Re-reeling**

This should be carried out on large reels with a circumference of 1.5 m and the standard re-reeling speed of 160 r.p.m.

Calculation of results

The results of at least two or three reeling tests must be taken in order to calculate the classification. The resultant raw silk weight divided by the sample cocoon weight will indicate the raw silk percentage. The cocoon classification items are worked out by the following method:

1. Percentage of cocoon shell

$$\text{Percentage of cocoon shell} = \frac{\text{Weight of cocoon shell (g)}}{\text{Weight of whole cocoon}} \times 100$$

Points to be observed for testing percentage of cocoon shell are as follows:

- The percentage of the cocoon shell can be calculated from the weight of the whole cocoon and cocoon shell obtained from 200 cocoons from which 100 cocoons are taken separately.
- If possible, equal amounts of female and male cocoons should be selected for the sample.
- No defective cocoons should be included in the sample.
- As atmospheric conditions make the moisture content of the cocoons too variable, the sample should be selected from a point 10 cm beneath the surface of the batch of cocoons.

The average value of the percentage of the cocoon shell obtained from repeating the test twice must be graded within a 0.3 percent deviation. But if the difference between both samples is + 0.3 percent another test should be taken.

2. Estimated cocoon percentage

$$= \frac{\text{Weight of eliminated cocoon}}{\text{Weight of sample cocoon (g)}} \times 100$$

The result should be expressed to one decimal place where the weight of sample cocoons is the sum of eliminated cocoons plus the weight of the good cocoons. Example:

$$\frac{\text{Weight of sample cocoons (dried) 1905 g}}{\text{Weight of eliminated cocoons (dried) 48 g}} = \frac{48\text{ (g)}}{1905\text{ (g)}} \times 100$$

3. Length of cocoon filament

$$\frac{\text{Raw silk length (m)} \times \text{Ave. reeling cocoon number per thread}}{\text{Total reeled cocoon number}}$$

Where, total reeled cocoon number

= Sample cocoon number – Converted carry over cocoon number

$$\text{Ave. reeling cocoon number per thread} = \frac{\text{Sum of reeling cocoon number per thread}}{\text{Sum of reeling ends number checked}}$$

Length of raw silk is checked by the gauge. Sum of reeling cocoon number per thread is the total number of reeling cocoon verified 20 times during the reeling (this means once per unit work process). The length is based on the average of every reeling block. It is expressed by total number to the one decimal place. Converted unreelable cocoon number and carry over cocoons are calculated by converting them into length and expressed by full cocoon number to the one decimal place.

The cocoon number converted to full cocoon length

= 1.00P + 0.77H + 0.39M + 0.12L (1)

Where P : number of newly cooked cocoons but unreelable
H : number of heavy shell cocoons but unreelable or carried over
M : number of middle shell cocoons unreelable or carried over
L : number of light shell cocoons unreelable or carried over

The reeling cocoon number per thread is estimated down to two decimal places. The reeling work is not completed to the last single cocoon, but up to about 50 cocoons: the remaining cocoons are "carry over cocoons". They are divided into three kings: (H)...cocoons where only the outside layer has been reeled, (M)...cocoons reeled up to middle layer, and (L)...cocoons reeled up to inner side layer. These cocoons can be reeled along with other cocoons. That is why they should be converted into full cocoon length or weight and then the converted number has to be deducted from the total sample of cocoons to obtain the exact raw silk percentage or actual sample.

The converted number of full cocoon number is called "carry over cocoon number" which is calculated by multiplying the number of cocoons with Heavy layer, Middle layer and Light layer by cocoon convert indices. Also, the unreelable cocoons during the reeling or after reeling, are converted into full cocoon number by applying the same indices as for carry over cocoon.

For example:

Number of sample cocoons	340
Length of raw silk	55 150 m
Total cocoon number of snap check on reeling cocoon number per thread	720
Total ends checked for reeling cocoon number per thread	91 ends
Unreelable cocoons	H(1), M(3)
Carry over cocoons	H95), M(14), L(21)

Reeling cocoon number = 340-1-12 = 326

Where (1 – Convert unreelable cocoon number)
(12 – Convert carry over cocoon number)

Ave. reeling cocoon number per thread = 720 + 91 = 7.91

$$\text{Length of cocoon filament} = \frac{55150\,(\text{m}) \times 7.91}{326} = 1\,338\,(\text{m})\ldots\text{batch(a)}$$

This is the length of cocoon filament in (b) batch is 1 349.1, the average of (a) and (b) batch is 1 343.6 m. Thus, the length of cocoon filament = 1 344 m.

4. Reelability percentage

$$= \frac{\text{Reeled cocoon number}}{\text{Number of ends feeding}} \times 100$$

Where reeled cocoon number = number of sample cocoon
- number of unreelable new cocoons
+ number of converted carry over cocoons

Number of ends feeding = number of cocoons fed
+ number of carry over cocoons
- number of converted carry over cocoons

Note: Reelability percentage means the average percentage of each reeling batch. It is estimated based on a total reduced to the decimal place. The converted carry over cocoons is shown by length level worked to one decimal place. Where, the reeling cocoon number and end feeding number are corrected by the conversion systems as follows:

Number of sample cocoon	340
Total of end feédings	521
Unreelable cocoon	H(1)
Carry over cocoons	H(5), M(14), L(21)

Reeled cocoon number	= 340 – 1 – 12 = 327
∴ End feeding number	= 521 + 40 – 12 = 530
∴ Reelability percetage	= (327 ÷ 530) x 100 = 61.69 batch (a)

If reelability percent of batch (b) is 60.99 percent the average of batch (a) and (b) will be 61.3 percent.

5. Raw silk percentage of cocoon

$$\frac{\text{condition silk weight (g)} + \text{carry over cocoon silk yield}}{\text{weight of cocoons}} \times 100$$

Now, converted carry over cocoon silk (g)
= weight of bave (g) x number of carried over cocoons to be converted to full ones.

$$\text{Here, weight of bave} = \frac{\text{conditioned silk weight (g)}}{\text{number of reeled cocoons}}$$

Where, number of reeled cocoons
= number of sample cocoons
- number of converted unreelable cocoons
+ number of converted carry over cocoons to full one

Note: Raw silk percentage is estimated to two decimal places. The weight of bave is estimated down to three decimal places. The converted unreelable cocoons and carry over cocoons are based on the weight system, and the number of converted cocoons is estimated on the total to one decimal place.

For example: The cocoon number converted to full cocoon weight
= 1.00P + 0.73H + 0.30M + 0.8L (2)

Where,
P – number of newly cooked cocoons, but unreeled
H – number of heavy shell cocoons, but unreelable or carried over
M – number of middle shell cocoons, but unreelable or carried over
L – number of light shell cocoons, but unreelable or carried over

Weight of cocoons sample – for example:	batch (a) : 300 g, batch (b) : 300 g
Conditioned silk weight:	batch (a) : 131.05 g, (b) : 131.75
Sample cocoon number:	batch (a) : 340, (b) : 340
Unreelable cocoons:	batch (a) : P(1), M(3)
	batch (b) : P(1), H(1), M(2), L(3)
Carry over cocoons:	batch (a) : H(5), M(14), L(21)
	batch (b) : H(3), M(11), L(27)

$$\therefore \text{reeled cocoon number} = 340 - 1 - 1 - 10 = 328 \text{ batch (a)}$$
$$340 - 1 - 1 - 8 = 330 \text{ batch (b)}$$

$$\text{Weight of bave} = (131.05\text{g} \div 328) = 40.0(\text{cg}) \text{ batch (a)}$$
$$(131.75\text{g} \div 330) = 39.9 \ (\text{cg}) \text{ batch (b)}$$

$$\text{Converted carry over cocoon silk yield} = 0.400 \text{ g x } 10 = 4\ 000 \text{ g batch (a)}$$
$$0.299 \text{ g x } 8 = 3\ 192 \text{ g batch (b)}$$

$$\therefore \text{Raw silk percentage of cocoon} = \frac{131.05 + 4.00}{300} \text{x} 100$$
$$= 19.267\% \text{ batch (a)}$$
$$\frac{131.75 + 3.192}{300} \text{ x } 42.8 = 19.252\% \text{ batch (b)}$$

The average of (a) and (b) batches is 19.26 percent.

Cocoon grading method

In cocoon classification, the result for length of cocoon filament and result of reelability percent is shown in Table 11 (1), (2) added up to the grading result which is applied to the cocoon grading shown in Table 11 (3).

Cocoon classification is divided into 5 grades: A, B, C, D and E.

Table 11. Cocoon classification

(1) Grading of cocoon filament length (m)

Length of cocoon filament	below 920	921 to 990	991 to 1060	1061 to 1130	1131 to 1200	1201 to 1270	1271 to 1340	1341 to 1410	1411 to 1480	1481 over
Mark	33.5	34.5	35.5	36.5	37.5	38.5	39.5	40.5	41.5	42.5

(2) Grading of cocoon reelability percent

Reelability percent											
	multi-end	below 39	40 to 45	46 to 51	52 to 56	57 to 62	63 to 68	69 to 73	74 to 80	81 to 86	over 87
	automatic	below 34	34 to 39	40 to 45	46 to 51	52 to 57	58 to 63	64 to 69	70 to 76	77 to 82	over 83
Mark		43.5	44.5	45.5	46.5	47.5	48.5	49.5	50.5	51.5	52.5

(3) Final grade ([1]) + [2])

Grade	A	B	C	D	E
Result	over 90	88-89	86-87	84-85	below 83

3.5 Cocoon exchange

The price of cocoons required for producing one kg of raw silk, (Cocoon Price Parameter [COO]), is calculated on the basis of 10 percent raw silk percentage cocoon to make 2A grade raw silk. The cocoon price parameter as a basis for estimating the cocoon price is obtained from the following formula:

$$\text{CPP/100 kg cocoon} = \frac{\text{(raw silk price + silk by-product income)} \\ \text{- (cost of raw silk production + profits}}{\text{1 bale weight of raw silk (60 kg)}}$$

As the CPP by this method stands as the standard of C grade cocoons, if varied from it each 2.0% CPP per single grade is added to or deducted from the standard one.

Grade	A	B	C	D	F
% of CPP amended	+4.0	+2.0	0	-2.0	-4.0

Cocoon is computed from the following formula

Cocoon price/kg = CPP ± amended CPP) x raw silk percentage of cocoon x (1-Discounting rate)

The resulting raw silk percentage is the raw silk test value multiplied by 0.9785, which equals fresh cocoon yield. The silk producer may have large stocks of cocoons in storage for reeling, which will result in the reduction of raw silk percentage. As the test sample does not match actual raw silk sold, consumed for testing in mills or in a "discounting rate" is used for raw silk during the reeling process of 2.15 percent (established in 1982).

Where no cocoon classification system exists, a derived arbitrary raw silk percentage is used based on visual inspection. A fictitious raw silk percentage is calculated using cocoon shell percentage, relative raw silk percentage from the previous year's assumed value. For example, in Table 12 when the CPP is 29 and the fictitious raw silk percentage of cocoon is 18 percent, the commercial price per kg is 29 x 0.18 = US$5.22.

Table 12. An example of cocoon pricing by visual inspection

Grades	Shell % of fresh cocoon	Percentage of good cocoon (%)	Fictitious raw silk % of fresh cocoon	Price of fresh cocoon ($ per kg)
1st	22.5 over	98	18	5.22
2nd	21.5	97	17	4.93
3rd	20.5	96	16	4.64
4th	19.5	95	15	4.35
5th	18.5	94	14	3.19
Fair cocoon		Outside slight stained	11	3.19
Undergrade cocoon			8	2.32

CHAPTER 4

COCOON DRYING, STORAGE AND SORTING

4.1 Objective of cocoon drying

The key goal of cocoon drying is to protect cocoon quality, preserve condition of cocoons for reeling and prevent damage caused by long periods of storage. The first hazard is the continued metamorphosis of the pupa. A newly emergent moth will pierce the shell rendering the cocoon useless for conversion to raw silk. Exposure to excessive moisture within the cocoon causes putrification and moulds. Drying kills the pupa and evaporates moisture that would, otherwise, ruin cocoons.

4.2 Mechanism of cocoon drying

Water content will vary according to silkworm type, rearing seasons and whether a cocoon is produced by a male or female. Table 13 shows the moisture content of the cocoon.

Table 13. Optimum moisture content of dried cocoons

Division	Fresh cocoon (%)	Dried cocoon (%)
Cocoon shell	11 – 12	6 – 7
Body of pupa	75 – 79	7 – 13
Whole of cocoon	61 – 54	8 - 12

As the majority of water is contained in the body of the pupa, the extent of drying will depend upon the actual moisture content. When fresh cocoons are placed in the drying machine, water in the cocoon shell evaporates rapidly and heat enters the pup through the shell. After the pupa is dead, generally 10 minutes water in the pupal's body evaporates very quickly. Cocoon drying continues gradually and when the desired amount of water has evaporated, the speed of drying is reduced relative to the decrease of moisture until drying is completed.

a) Drying temperature

The drying temperature impacts the cocoon shell and its resultant raw silk yield. For example, if the temperature exceeds certain limits, sericin degenerates and concurrently efficiency, reelability and raw silk percentage declines. It is advisable to observe the following limits: For hot air drying 115 ± 5°C is recommended on contact with the cocoon. For steam-heat drying 102 ± 2°C is the preferred guideline. While setting a higher finishing temperature increases drying efficiency, the melting point of sericin declines when the drying rate exceeds 50 percent. The finishing temperature should be reduced gradually from 60°C in hot air drying and 55°C in steam-heat drying.

b) *Effect of wind humidity and velocity*

The humidity of drying air has little influence on cocoon quality: however poor ventilation can cause high temperature and humidity, which will hurt reelability quality. It is advisable to maintain the relative humidity at 4-5 percent at the initial stage of drying, and at 16-19 percent at the later stage, to prevent excessive reductions of water in the cocoon shell.

Wind velocity as an isolated factory has little effect on the cocoon quality. However if air pressure is not uniform throughout in the drying machine there is a risk of uneven drying.

4.3 Various methods of stifling/drying

In countries like Japan that have advanced techniques of sericulture, bivoltine cocoons are dried using hot air generated by electricity or steam. This modern method is only suitable for bivoltine species. In tropical areas where multivoltine cocoons predominate, the main procedure is steam stifling. The following describes the three methods of cocoon drying in commercial use:

a) *Sun drying*

No investment is required to kill pupae and dry cocoons in bright sunshine. Clearly, this is only possible in tropical and sub-tropical zones. Fresh cocoons are spread in thin layers on a mot or planks of wood and exposed to direct sunlight. Depending on the strength of the sun, the process takes two to three days. Though cheap and simple to employ, the main disadvantage is silk fibre's sensitivity to ultra-violet rays, which harm fibre strength and colour. Since there are limited facilities for quick marketing of cocoons, sun drying continues to be utilized in many tropical and sub-tropical countries.

b) *Steam stifling*

Generally, in many tropical and sub-tropical countries, fresh cocoons are heaped into a basket and steamed in small reeling units. In big reeling units with a boiler, the steam chambers are used for stifling. Chamber steaming represents an improvement over baskets as the cocoons are spread in thin layers. Steam works well for multivoltine cocoons, which are soft and reel easily without long periods in storage. After steaming, these cocoons may be easily cooked in an open pan, a method commonly used in India. Steam is rapid as pupae are killed within 30 minutes. Immediately after stifling, the cocoons are spread on spacious, well-ventilated shelves and left for three to four days partial drying prior to reeling. The cocoons must be turned over frequently to prevent the growth of mould. If left for extended periods, such as more than a month, the risk of mould is pronounced even with frequent turning and satisfactory ventilation. This damage is especially acute during the rainy season.

c) Hot air-drying

The hot air drying method is very common for bivoltine cocoons in sericulturally advanced countries. The hot air dryer essentially comprises:

i. Drying chamber, in which fresh cocoons are placed preferably in thin layers

ii. A fan to maintain constant and uniform air current throughout the layers along with an efficient ventilator to drive out moisture avoiding condensation inside the chamber, and

iii. A heater for heating the air driven by the fan. Precise thermostatic control for regulating the temperature should also be installed in the drying.

4.4 Degree of drying

Optimum percentage of drying signifies that fresh cocoons have been processed from 37 to 42 percent to withstand storage over a long period. Excessive drying occurs when percentages exceed the ideal range. For example, 40 percent of drying means that 100 g of fresh cocoons have been dried up to 40 g. Correct drying enables cocoons to withstand long-term storage of 6 to 12 months without developing mould. When the fingers can crush the pupa, the optimum degree of drying has been achieved. Much skill is needed to derive this ratio empirically. One way is to add a factor of 20 to the cocoon shell percent. A new way is application of the following formula for optimum drying.

Optimum percentage of drying = (0.0115 x mc of fresh cocoons – 0.2104)
x percent of cocoon shell – 1.15
x mc of fresh cocoon + 115

(mc = moisture content)

4.5 Types of drying machines and methods

The silk industry now utilizes a range of drying machines and attachments suited to available technology plus drying conditions. The following are most popularly used: shelf-carrier/cabinet type, one band type, multi-step and low temperature air duct type.

a) Shelf-carrier/Cabinet type dryer

This type performs the drying operation through a chamber, which has layers of trays with perforated bottoms. Numerous steam pipes are laid in parallel under the bottom and on opposite walls to heat the inside of the chamber. Fans are fixed on opposite walls in order to produce the parallel air current on each tray of the shelf-carrier. Radiated heat and natural air currents (Figure 5) then dry the cocoons.

Drying method: Initial temperature is 100°C ± 2°C, which is gradually lowered to the finishing temperature of 60-50°C.

The drying time from fresh cocoons to the required dried state is about 7 hours. Because of its simple structure, this type of dryer can malfunction if temperature control and distribution is not skilfully carried out. Considerable manual labour is needed to spread cocoons on the trays and remove them after completion. The Cabinet type for cocoon testing places the fresh cocoons in drawers for drying and is equipped with an aerofin heater and powerful electric fan at the upper section. The electric fan makes the air circulate inside the machine by changing the revolution direction at given intervals for minimizing uneven drying. This style is adequate for the pilot silk reeling plants or research institutes.

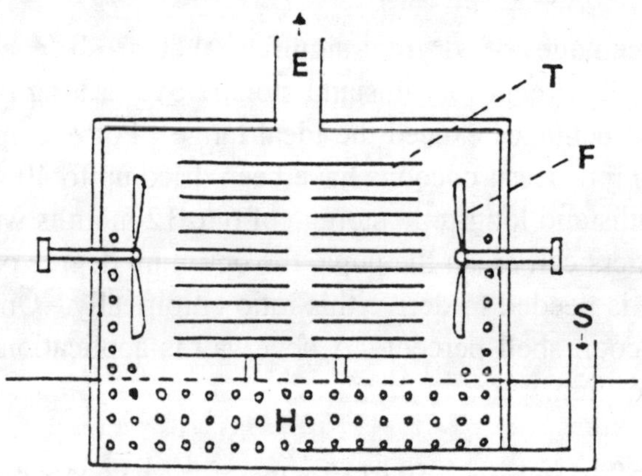

Figure 5. Shelf-carrier/Cabinet type dryer
T: cocoon tray, H: heater, F: fan, S: air inlet, E: air outlet

b) One band type dryer

This is a one-band type continuous system medium-sized cocoon drying machine incorporating wire netting conveyers. Cocoons are spread on the wired net conveyor at 300 to 400 m/m thick by a cocoon supplying conveyor. The machine consists of four to eight chambers according to the drying capacity with a moving conveyor and equipped with an air blower, heater, air supplying and exhausting dampers and thermo controller in each chamber.

In addition, it has an air cooling chamber where the dried cocoons are cooled (Figure 6). Hot air passes through the heater permeating into the layers of cocoons from the top to the bottom, and vice versa, by air changing over damper which connects with a timer which automatically opens and closes. A part of the air is extinguished, while fresh air is sucked from outside into the air blower for circulation. This type has strong advantages. It takes up little space and each section of the dryer has an exact drying temperature. However, heavy electric power and much skill are needed in order to achieve even drying, compared to the multi-step hot air dryer.

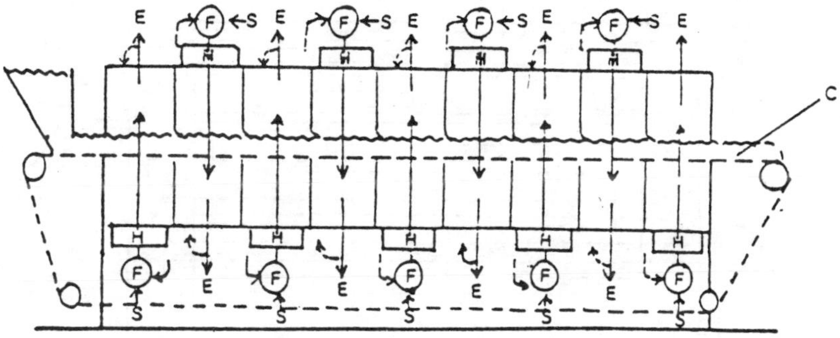

Figure 6. One step band type dryer
F: fan, H: heater, S: air inlet, E: air outlet

c) Hot air circulating type dryer (multi-step)

This machine dries cocoons with hot air heated outside of the machine, which is then blown in. It has three sections: upper, middle and lower, which have different air heating conditions (Figure 7 and Figure 8). The multi-step improves drying efficiencies by being able to maintain the drying conditions acceptable by perpetuating a steady circulation of air.

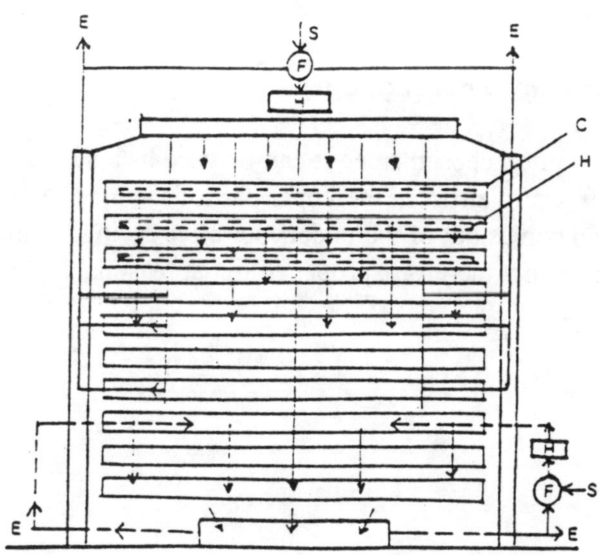

Figure 7. Hot air circulating type dryer (section)

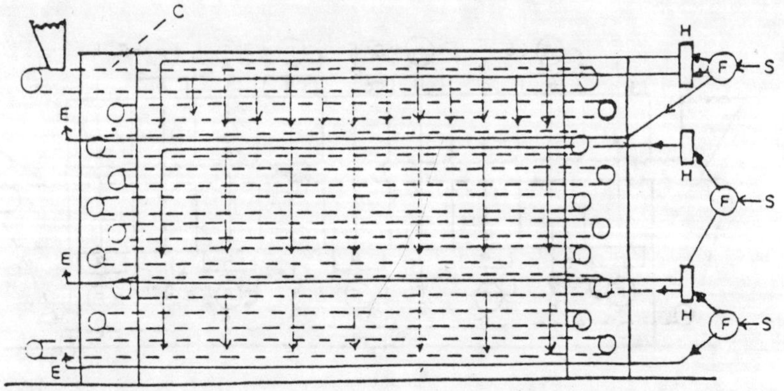

Figure 8. Hot air circulating type dryer (side view)
C: conveyor, H: heater, F: fan, S: air inlet, E: air outlet

The temperature at the commencement of drying is 110-115°C, which is then lowered in stages to 100, 90, 80 and 70°C until the final temperature of 60-55°C is reached. The duration of drying from fresh cocoon to standard dried condition is 5.5 to 6.0 hours, when the target degree of 39-42 percent drying can be reached. There are various kinds of this type of dryer, such as the Six-step, Eight-step and Ten-step. The Eight-step temperature dryer is the most preferred. This type has strong advantages such as even drying achieved by a proper mixing of cocoons and economical heat energy settings for drying, even though it is expensive and demands much space for installation. This dryer is, therefore, suitable for a large-scale modern silk reeling factory.

d) Low temperature air duct type dryer

This type is often chosen since it can be used as a storage room after use. As shown in Figure 9, warm air from the upper part of the air chamber goes through the cocoon storage and comes out of the underfloor after drying. Thick layers of cocoons can cause uneven drying: therefore, proper filling at reasonable intervals is desired.

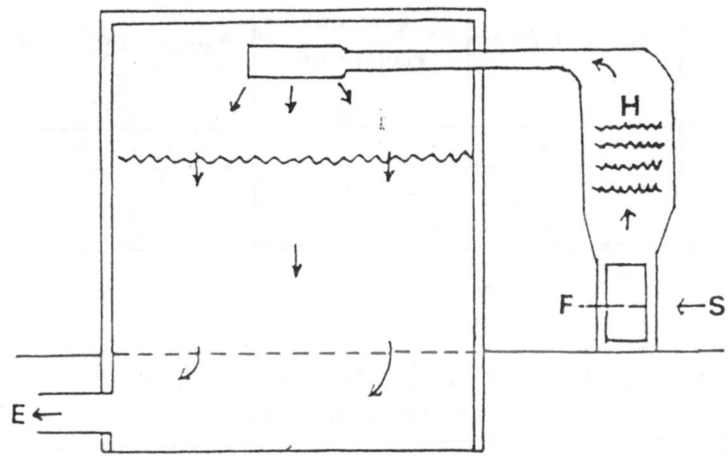

Figure 9. Low temperature air duct type dryer
F: fan, H: heater, S: air inlet, E: air outlet

The fresh cocoons for drying are piled and spread uniformly inside the drying room at a thickness of about 15 cm; the hot air at 80°C blows in for 4 hours. This work can be repeated until the filling of the cocoons reaches the desired height. After this, the hot air, which is gradually lowered by 5°C at intervals of 4 hours, should be blown in until it reaches the same temperature outside the drying room.

This method has strong advantages such as fewer installation costs and the dual purpose of drying and storage if continuous supplies of fresh cocoons are obtained during the drying work. There are, however, some constraints to this method. It is difficult to check the drying degree of cocoons; and there is a likelihood that the Cleanness of the raw silk quality deteriorates due to the improper swelling of cocoons during the cooking process. Therefore, greater caution should be exercised when handling this dryer.

4.6 Effects of drying conditions on reeling results

The initial drying temperature has the largest effect on the cocoon shell. When the temperature exceeds the highest limits, sericin is sharply degenerated leading to a decrease in raw silk percentage. In the case of steam-heat radiation drying, if the temperature is raised too high, it is liable to cause uneven drying as radiated heat degenerates the quality of the cocoon, bringing about bad results. If the initial temperature for cocoon drying is lowered too much, it is apt to deteriorate the neatness and cleanness result of the raw silk quality (Table 14). Incomplete drying may cause uneven cooking results and consequently lower the raw silk quality in relation to cleanness and neatness (Table 15).

Table 14. Initial temperature for cocoon drying (Choe, *et al.* 1971)

Drying temperature	Raw silk (%)	Cocoon reelability (%)	Neatness (%)	Cleanness (%)
98-65°C	17.16	62.7	91.4	92.8
110-65°C	16.95	55.6	91.5	92.2
75-55°C	17.07	62.1	88.1	90.0

Table 15. Effect of incomplete drying on result of reeling (Choe, *et al.* 1971)

Treatment	Raw silk (%)	Cocoon reelability (%)	Neatness (%)	Cleanness (%)	Remarks
Optimum drying	16.96	59.5	91.3	94.7	Drying ratio : 42%
Incomplete drying	17.16	60.5	89.3	93.1	Drying ratio : 44%

4.7 Cocoon storage

For dried cocoon storage, the best conditions will keep the raw materials for long periods without any damage from moulds and pests. In building the cocoon storage room, the following should be considered. Air space for ventilation must be provided at the bottom of the storage room, to prevent moisture coming up from underground as shown in the Appendix, Figure 9.

The cocoon stores should preferably be built with double walls.

If there is no double wall, every endeavour should be made to have the walls as thick as possible. Windows should be small and fixed at a high level and provided with an exhaust fan to enable moisture accumulating in the room to be drawn out. The storing room should be lined with an iron sheet, the windows should be covered with a wire net of about 1.5 mm mesh, the entire interior of the storeroom should be lined with wood panelling, including all walls, ceiling and floor.

Cloth or polyethylene bags are recommended as containers for dried cocoon storage. Cocoons weighing 200 litres are generally put in a bag. Shelves should be provided in three to four tiers and cocoon bags should be stored there. Bags should not be piled together. Alternatively, the warehouse should be partitioned into many small rooms, in which dried cocoons can be kept loosely. A cocoon storage bin "Honey Comb Tub Type" is recommended to reduce the fungus damage caused to dry cocoons during storage, especially during humid seasons (Appendix, Figure 8).

If humidity exceeds 70 percent, action should be taken to reduce humidity such as scattering calcium chloride ($CaCl_2$) or silica gel on the floor. (Chemicals can be recycled by heating.) Also, a heater can be used. Precautions should be taken to prevent fires.

After a long period of rain, if it is found that the pupa on testing has a moisture content of 15 percent or over, it will be necessary for the cocoons to be taken out on a clear day and spread out for drying in the shade.

If there is danger of fungal damages, it will be advisable to redry the cocoons at 60°C for two hours.

4.8 Sorting of cocoons

Silkworm farmers remove defective cocoons as well as the double cocoons before taking the produce to the market. Still there may be a small percentage of defective cocoons, which would have escaped preliminary sorting. Hence the cocoons require a second sorting to ensure uniform good quality cocoons for reeling. The type of defect generally encountered in the second sorting are double cocoons, crushed and stained cocoons, cocoons with prints of cocooning frame, flimsy cocoons and insect damaged cocoons, thin shelled cocoons and mouldy cocoons.

a) Method of sorting

The cocoons are spread out on tables with low partitions and the sorters sit around the tables and pick out the defective and double cocoons by visual inspection. The sorting room is generally located close to the cocoon stores and is provided with good ventilation and lighting.

Rejections are placed in a separate container. In the procedure described above, only obviously defective cocoons are rejected. This system of sorting, therefore, does not commend itself to modern techniques of reeling which aim at producing predetermined qualities of raw silk.

In advanced countries, the sorters detect even internally damaged cocoons and eliminate them by passing the cocoons over ground-glass plates illuminated from below. These glass plates generally measure about 38 x 38 cm^2 and are fitted in cut-outs of equal size in a lamp blacked table top or moving platform (Appendix, Table 13).

A fluorescent light or sometimes even an ordinary 60-watt frost lamp placed in a suitable position under the platform provides the required illumination from below. When defective cocoons containing decomposed pupa and those with the inside discoloured pass over the glass plate they are easily detected. This method of sorting not only enables thorough sorting but is also economical, since the sorting work is made easier for the workers.

b) Cocoon mixing

In most modern factories, which aim at producing high-grade raw silk, the kinds of cocoons graded in the visual inspection or mechanical test are actually mixed in required proportions. This is called cocoon mixing or blending and is done to ensure speed and uniformity of reeling as well as to obtain the desired effect in raw silk. This sort of blending of cocoons has been found to be quite essential for ensuring a high degree of efficiency of automatic reeling machines.

Table 16. Qualities of raw silk reeled from sorted cocoons (Song and Choe, 1970)

Items	Cocoons with decomposed pupa	Inside stained cocoons	Fluffy cocoons	Outside stained cocoons	Thin shell cocoons	Frame printed cocoons	Good cocoon
Cleanness (%)	74.8	83.7	85.8	85.1	84.5	93.4	95
Average neatness (%)	76.0	88.7	80.6	86.7	87.0	89.4	91
Low neatness (%)	69.3	78.8	73.3	78.5	80.0	82.3	83
Grade	E	D	E	C	D	A	2A

CHAPTER 5

COCOON COOKING

5.1 Introduction

Cocoon cooking unwinds the cocoon filament spun by the silkworm. The sericin covering around the cocoon filament is agglutinated after silkworm spinning, then hardened through the cocoon drying process. In preparation for reeling, it should be softened.

Processing softens sericin by heat, water and steam. Ideally there will be uniform softening of the outer and inner cocoon shell.

5.2 Cocoon cooking methods

1. Pan cooking

a) Cocoons are put into a pan of boiling water. The groping ends of the cooked cocoon can be completed with the stirring rod in the pan (see Appendix, Figure 14).

b) The cocoons contained in the wire cage are placed into the boiling water and then boiled for a few minutes. After boiling, the cocoons with the wire cage are moved into a low temperature bath in another pan. If there is water permeation inside cocoon shell and swelling of the cocoon shell, this work can be repeated (see Appendix, Figure 15).

2. Machine cooking

Recently, machine cooking has become widely used in most silk reeling factories (Figure 12). In general, the machine cooking process is divided into six parts as follows:

a) Soaking part

The surface layer of the cocoon swells over at the soaking part. The first part is carried out by dipping the outer layer of the cocoons into a water bath at 55°C. This step is comparable to fabric dipping for dyeing to ensure even results.

b) High temperature and low temperature permeation part

The uniform cooking of the outer and inner layers of the cocoon can be easily attained by replacing the air of the cocoon cavity with water since water transfers heat faster than air. Thus, the wet cocoons from the soaking part are exposed to steam at about 90-95°C at an appropriate steam pressure. Next, the air inside the cocoon cavity is heated. When cocoons treated at high temperature permeation are moved into low temperature permeation around 65°C, partial condensation occurs in the cocoon cavity. Then the cocoon sucks water in, evenly wetting all the layers of the shell.

The amount of water permeated into the cocoon cavity is controlled by the difference in temperature between high temperature and low temperature permeation parts, as well as the air permeability of the cocoon shell.

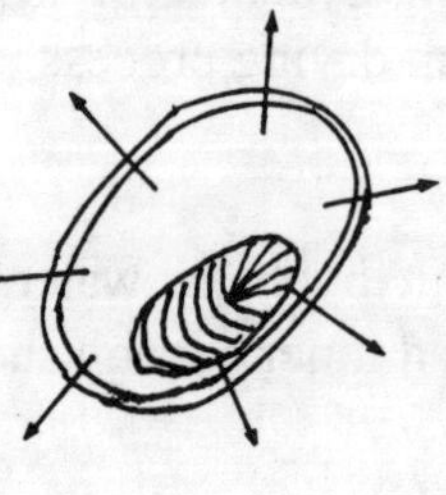

Figure 10. Air expansion in cocoon cavity and partially replaced with steam at high temperature permeation part.

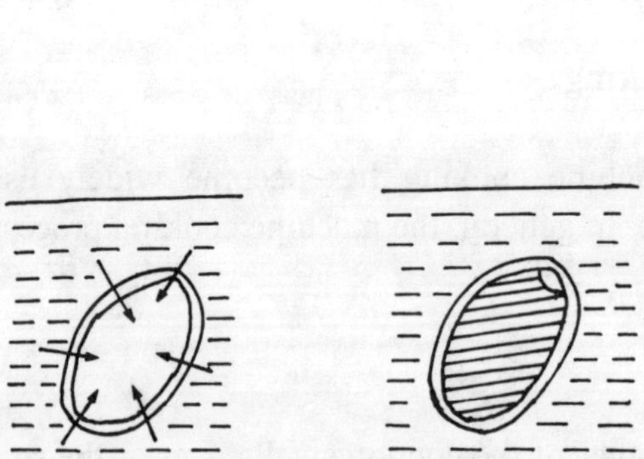

Figure 11. Cocoon sucks in water at low temperature permeation part

c) ***Steam cooking part***

Now the cocoons treated in steps *a)* and *b)* are moved to the steam processing. This part causes the sericin to swell and soften the silk layers and the steam to fill up the cocoon cavity by diffusing the permeated water out of the cocoon. Steam heat is a highly efficient head conductor that can cause unwanted sericin loss. To minimize or stop this loss of sericin, infra-read rays have been attempted in this process, but proven unfeasible. For cocoons anticipated having poor reelability, the steam cooking part has to be prolonged.

Sudden variations in steam pressure can adversely affect the cooking process by producing over processed or insufficiently processed cocoons. These poorly cooked cocoons seriously deteriorate reeling efficiency by decreasing raw silk yield and quality and boosting cleanness defects during reeling. It is urged that greater vigilance be exercised to control steam cooking.

d) ***Cooking adjustment part***

At this point the steam content of the cocoon cavity is replaced with water through gradual condensation of steam in the cocoon. This is effected by gradually cooling of the water from 98° to 65°C. Sericin swollen by steam cooking becomes stable. As this step consumes large volumes of water for cocoon permeation, it needs to be longer in duration and requires more fresh water than other parts.

e) ***Low temperature-finishing part***

Cocoons are finished in 50-60°C water. Here, the cocoons, which were properly swollen by processing in the first five steps, become more stable and are prepared for the next stage in reeling.

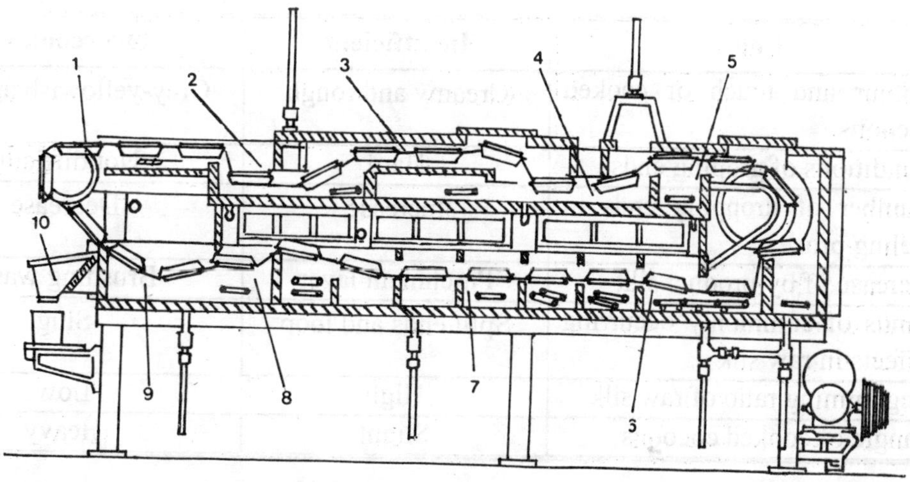

Figure 12. Cocoon Cooking Machine
1. Cocoon hopper part, 2. Soaking part, 3. High temperature permeating part, 4. low temperature permeating part, 5. Cooking part, 6-7. Adjusting part, 8-9. Low temperature finishing part, 10. Cocoon outlet

5.3 Degree of cocoon cooking

Optimum degree of cocoon cooking can be gauged correctly by the reeling results, but it also can be evaluated as follows:

a) Response to touch: In the case of optimum cooking, the cooked cocoon is velvety and the shell is neither soft nor hard when felt by the hand. In the case of over cooking, the cocoon is soft and collapses easily. under finger pressure. If under or imperfectly cooked, it is hard.

b) Groping and condition: In case if optimum cooking, the filament unwinds by itself at the moment the groped end is picked up and grasped by hand, with little resistance on the fingertip, and without breaking the end.

c) Cocoon colour: In imperfect cooking, it is creamy and cooked partially. When over cooked, the cocoon is grey-yellow.

d) Cooked cocoon shell weight: five to six times as heavy as the original shell weight is desirable; 4.5 times or less is defined as under or imperfect cooking, while the overcooked cocoon weighs six times or more than the original.

e) Cooked cocoon weight: 10-11 times the weight of the original dried cocoon.

f) Conditions of cooked cocoons in basket: The cooked cocoons should easily drop out of the basket, when the baskets are opened at the end of the cooking process.

g) Crushed cocoon: There are no crushed cocoons in optimum cooking.

Table 16a. Criteria of insufficient and overcooking

Items	Insufficient	Overcooking
Colour and touch of cooked cocoons	Creamy and rough	Grey-yellowish and soft
Conditions of groping ends	Hard	Not difficult
Number of dropping ends in reeling process	Increase	Decrease
Increase of by-product silk	Parchment layer	Brushing waste
Kinds of commonly occurring defects in raw silk	Split ends and loop	Slug
Degumming ratio of raw silk	High	Low
Tangle of cooked cocoons	Slight	Heavy

5.4 Adjustments to cooking conditions

Cooking methods should be re-adjusted for better results depending on the different quality of cocoons as shown in Table 17.

During treatment, the temperature variations should be less for good reelable cocoons than for poor reelable ones, where steaming pressure and temperature of steam cooking should be increased. The temperature in cooking adjustment should be precisely maintained for good reelability cocoons. This level of precision is unnecessary for cocoons with poor reelability. Remedies for poor cooking results are shown in Table 18.

Table 17. Adjustment of cooking methods depending on variations in cocoon quality

Different cocoon qualities	Re-adjustment of cooking methods
Thick shell cocoon with high reelability	• In pre-treatment (soaking, steaming, permeation) difference of temperature should be more, while steam pressure and temperature of steam cooking part should be lower
Thin shell cocoon with high reelability	• Different of temperature part should be higher and steam pressure and temperature of steam cooking part lower.
Thick shell cocoon with poor reelability	• Difference of temperature in pre-treatment as well as steam pressure and temperature of steam cooking part should be more.
Insufficiently dried cocoons	• Steam pressure and temperature of steam cooking part less
Over dried cocoons	• Steam pressure and temperature of steam cooking part more

Table 18. Points for cooking operations

Phenomena	Counter measures
Crushed cocoons	• Reduce the difference of temperature between high and low temperature permeation parts as well as steam cooking part and first cooking adjustment
Overcooking	• Lower the temperature of steam cooking and high temperature permeation parts • Shorten the time for cooking • Modify the quality of reeling water
Insufficient cooking	• Increase gradually the temperature of steam cooking part • Prolong the time for cooking • Use the soluble type water for cooking instead of hard type one
Floating cocoons	• Make the temperature curve in cooking adjustment • Increase the difference eof temperature between high and low temperature permeation parts • Reduce the temperature of low temperature finishing part
Uneven cooking	• Increase the temperature of soaking part • Check the steaming pipe in the high temperature permeation part • Strict assorting of inferior cocoons

5.5 Effects of cocoon cooking conditions on reeling result

The suitability of cocoon cooking has an important bearing on the reeling result. A high proportion of waste is caused whenever cocoons are over- or undercooked. For overcooked cocoons, this happens when the filament comes out in lumps, clogs the reeling button and increases the number of breaks during reeling. Each time a break is repaired, the silk filament must be pulled out and all of the silk removed becomes waste. The undercooked filament does not unravel easily and also causes breakages and waste. In each case reeling efficiency is diminished.

Table 19. Effects of cooking degrees on reeling results (Lee, *et al.* 1971)

Degrees of cooking	Raw silk % of cocoon	Reeling troubles per 10,000 m (times)	Percentage of by-product silk		Neatness (%)	Cleanness (%)
Under cooking	16.00	1.47	1.28	1.23	92	91
Optimum cooking	15.98	1.60	1.45	1.10	94	95
Over cooking	15.94	1.67	1.54	1.03	92	93

CHAPTER 6

RAW SILK REELING

6.1 Introduction

Silk reeling is the process by which a number of cocoon baves are reeled together to produce a single thread. This is achieved by unwinding filaments collectively from a group of cooked cocoons at one end in a warm water bath and winding the resultant thread onto a fast moving reel. Raw silk reeling may be classified by direct reeling method on a standard sized reel, indirect method of reeling on small reels, and the transfer of reeled silk from small reels onto standard sized reels on a re-reeling machine. The last technique is primarily applied in modern silk reeling processes.

6.2 Various silk reeling devices

There are many types of silk reeling machines in use. The major structural features of the Sitting Type Reeling Machine, the Multi-ends Reeling Machine and the Automatic Reeling Machine are shown in Table 20.

1. Hand spinning wheel

This primitive spinning apparatus is operated by two hands – one to drive the wheel and the other to feed in cocoons. One end of the reeling thread is wound onto each wheel, while cocoons are boiled in a separate pot.

2. Charka type reeling machine

The Charka type is in use in India. This machine is operated with separate work motions in reel driving and cocoon feeding to reeling ends by two men per machine. Each machine has 3 ends or more to a reel, which is the same size as the large wheel of the Re-reeling machine in order to save the re-reeling process (direct reeling method).

3. Sitting type reeling machine

There are two kinds of sitting type reeling machines, foot operated and motor-driven (see Appendix, Figure 17). The motor-driven reeling machine is not equipped with the stop motion attachment (see appendix, Figure 18). There are obstacles to the production of good quality raw silk as the raw silk thread is wound too rapidly to maintain good quality control.

Table 20. Comparisons of structural features of various reeling machines

Kinds of machine	Sitting type reeling (m/c)	Multi-ends reeling (m/c)	Automatic reeling (m/c)
No. of reeling ends per basin	4 (2 to 8)	20 (10 to 40)	20
End-groping apparatus	Hand-driven	Semi-automatic	Automatic
End-picking apparatus	Hand-driven	Hand-driven	Automatic
Cocoon supplying apparatus	None	None	Equipped
End-feeding apparatus	None	Equipped (hand feeding)	Equipped (machine feeding)
Stop motion	None	Equipped	Equipped
Traverse guider	Equipped	Equipped	Equipped
Temperature of reeling bath	65-80°C	30-45°C	30-45°C
Reeling velocity (meter per minute)	180-250	50-80	120-200

4. Multi-ends reeling machine

This machine eliminates the disadvantages of the Sitting type reeling machine by increasing the number of reeling thread ends per basin and reducing the reeling speed. The operator must stand when running this machine as the number of reeling threads per basin increases by twenty-fold. This is also called a "Standing type reeling machine". Reeling efficiency is unchanged. Quality is better due to reduced speed.

The Multi-ends reeling machine is composed of driving part, groping ends, picking ends, standby bath, reeling part, jetboute, stop motion, traverse guider, small reels, steam heating pipes and clutches (see Appendix, Figure 19).

The cooked cocoons contained in the tubs are carried into the groping ends portion of the reeling machine. From there, cocoons are moved into the picking ends apparatus. After correctly processing, the cocoons go to the standby bath for cocoon feeding. They are picked up by the reeler and fed to the reeling thread. During this step a number of cocoons will be dropped thus reducing the ratio of reeling cocoons per thread. The normal speed of cocoon feeding by a skilled reeler is around 16 times per minute. The reeling thread passes through the jetboute, silk button, first guider, second guider, third guider, fourth guider, traverse guider, in that order and then is wound onto the small reels.

The cocoons dropped during the reeling process are gathered and reprocessed starting from the groping end section. The croissure of reeling thread is made between second guider and third guider, and the length of croissure is not for twisting of thread but for cohesion of thread by rubbing of composed filament. Typically, one set of Multi-ends reeling machines consists of ten basins with each basin having twenty ends or reels.

Basin: The basin is rectangular with well-rounded corners and edges. It is only 10 to 12 cm deep. It is commonly made of dark coloured porcelain. The basin is subdivided into sections, each intended for a specific job such as brushing, end gathering of baves, stocks in reserve and waste collection.

Reels: The reels of the Multi-ends reeling machine have a circumference of 75 cm. The frame of the reel is made of light metal or plastic. The reels are fitted into reel carriers and driven by a transmission shaft by connecting gears.

Traverse guider: To ensure narrow and long web on the hank of the reel, a cam type traverse assembly has been fixed. This will make a convex surface in the hank, which is wound on the reel. The centre part of the hank is higher than the two axis.

Thread button: Porcelain button thread-guiders are used for removing any dirt adhering to the thread passing through the tiny aperture in the button.

5. Automatic reeling machine

In raw silk production, the continuing increase of labour costs has mandated automation. Around 1950, the Automatic reeling machine, which controls the number of reeling cocoons per thread, was invented. Shortly thereafter, it was replaced by a second Automatic reeling machine, which could automatically control the size of the reeling thread.

The Automatic reeling machine mechanizes the processes of groping ends, picking ends; cocoon feeding to reeling thread and separation of dropped end cocoons during the reeling process. The efficiency of the Automatic reeling machine compares favourable with the manual Mult-ends reeling machine.

The Automatic reeling machine though built to replace manual reeling, still requires manpower for problems with the reeling thread, which must be corrected by hand. A moderate amount of cooked cocoons are carried to the newly cooked cocoon feeder (3) and then removed into the groping end part (4)

The end groped cocoons go to the picking end part (5) and the correctly picked end cocoons are dispensed to the cocoon supplying basket (8) which continuously rotates around the reeling basin on an endless chain belt. Usually, the reeling method is classified into the fixed cocoon feeding system and moving cocoon-feeding system.

In the case of the fixed cocoon feeding system, the correctly picked end cocoons in the rotating cocoon baskets are poured into the arranging basin (11) and here the picked end of each cocoon is hung on the end holding reel (7). When the size detector of the reeling thread indicates the feeding of cocoons, the picked end cocoons on standby are fed to the reeling thread by a feeding spoon. The reeling thread fed by picked end cocoons passes through the jetboute (23), silk button, first guider, second guider, third guider, fourth guider, denier indicator (5'), fifth guider and traverser (3'), and then it is finally wound onto small reels (1'). The end dropped cocoons are placed into the cocoon flowing tunnel by the remover plate. They are carried into the pupa separating drum (14). However, more reelable cocoons are poured into the end groping part (4) by the conveyor belt (15) and reels-finished cocoons are placed into the dropped-pupa case (17) for parchment layer cocoons (Figures 13 and 14).

In the case of the moving cocoon feeding systems, the correctly picked end cocoons are contained in the moving cocoon basket equipped with cocoon feeding apparatus. They are fed by the feeding fork of the cocoon basket, which move simultaneously around the reeling basin. The denier indicator of the reeling thread indicates the feeding motion of the cocoon. After cocoon feeding, the reeling path of the moving cocoon feeding system is the same as that of the fixed cocoon feeding system.

Generally, one set of the Automatic reeling machines has 400 ends, while one basin has 20 ends. The operating efficiency of the Automatic reeling machine is easily affected by cocoon qualities, drying and cooking machinery and quality of reeling water.

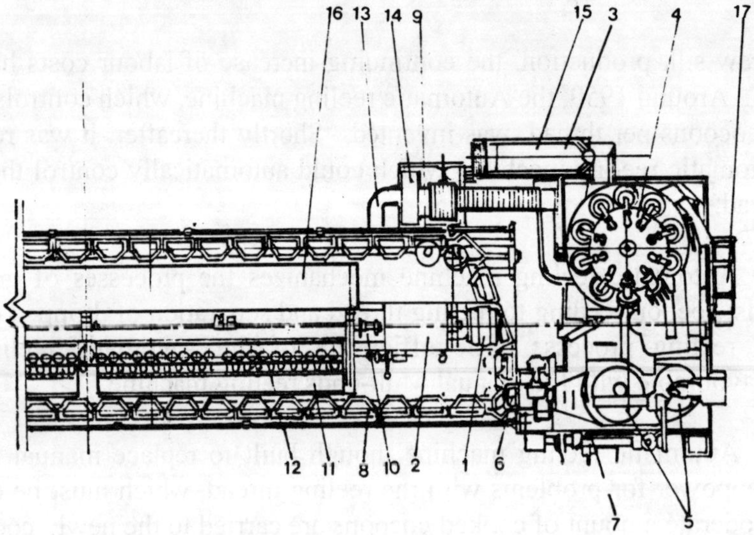

Figure 13. Ground plan of automatic reeling machine

1. **Motor,**
2. **Small reel speeding changing device,**
3. **New cocoon feeding device,**
4. **Brushing device,**
5. **Automatic end picking device,**
6. **Picked cocoon replenisher,**
7. **End holding reel,**
8. **Cocoon supplying basket,**
9. **Water supplying tank for basket,**
10. **Traverse gear,**
11. **Arranging basin,**
12. **Reeling basin,.**
13. **Conveying gutter,**
14. **Pupa separating device,**
15. **Conveyor belt case,**
16. **Roof cover,**
17. **Dropped-pupa case.**

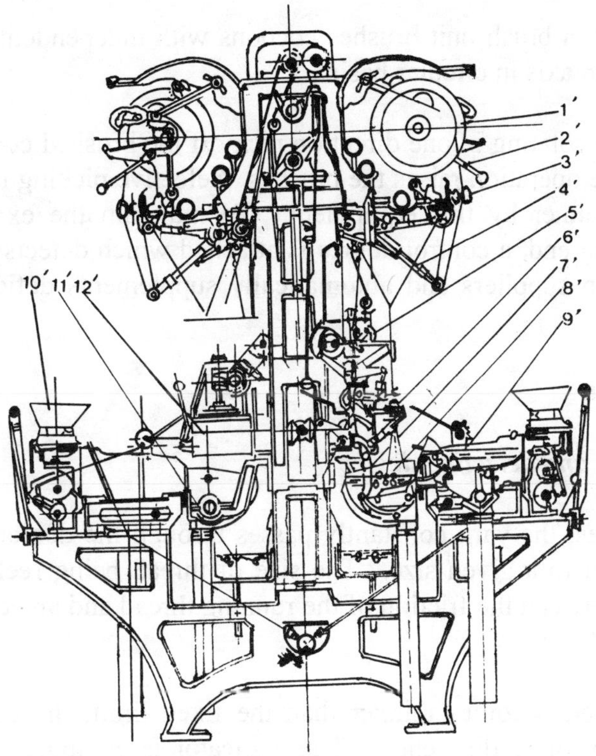

Figure 14. Side view of automatic reeling section
1'. Small reel, 2'. Rotating device, 3'. Traverser, 4'. Stop motion, 5'. Denier indicator, 6'. Jetboute, 7'. Shatter for reeling cocoons, 8'. Reeling basin, 9'. Arranging basin, 10'. Cocoon supplying basket, 11'. Conveyor gutter, 12'. Rotating device for jetboute.

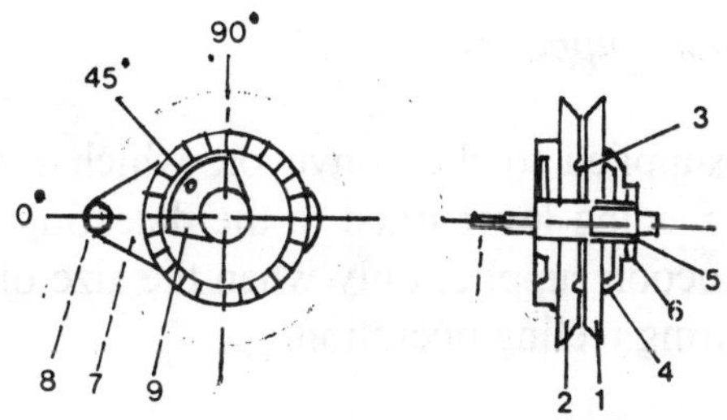

Figure 15. Size detector (gauge type)
1. Gauge glass, 2. Grooved gauge glass, 3. Gauge plate, 4. Washer, 5. Binding bolt, 6. Special nut, 7. Indicating ring, 8. Detection limiting rod, 9. Adjusting weight

1. Brushing and picking section

An automatic ten-brush unit brushes cocoons with independent arms rotating in reverse motion on an axis in circular basins.

Picking frames rotating in one direction pick off the brushed cocoon filaments when a cam during the operation raises the brushes. Selective picking is completed in the most effective manner by this equipment. To maintain the exact number of cocoons at each reeling end, a control device is attached which detects the amount of cocoons in the cocoon suppliers and automatically supplements deficiencies in the number of cocoons.

2. Reeling section

The denier indicating and detecting device

In these devices, the yarn constantly passes through the denier indicator and detector, which are set to a given size. The size of thread being reeled is detected through the balance between the friction of the running thread and an eccentric weight fitted on the denier indicator.

When the thread becomes thinner than the fixed limit, the denier indicator indicates the necessity of feeding-ends. This indicator is composed of two gauge glass plates with other pieces of gauge plate which have a slit between them corresponding to the given size of thread to be reeled. The size of yarn may be adjusted to the required sizes by varying the irregular weight with the denier-adjusting device; if a wider range of adjustment is required the denier indicators have to be replaced.

Conveyor system for cocoon suppliers

The fixed end feeding system is employed together with a conveyor system to carry cocoon supplies and feed cocoons to the reeling section whenever required.

End feeding for cocoons suppliers

Cocoons are supplied to the conveyor, which rotates constantly around the reeling section. The feeding lever fixed to the detecting mechanism will trigger the driving lever on the cocoon supplier only when the size of yarn becomes thinner than the required denier during reeling operation.

Stop motion

If there is a defect in the reeling, the reel is automatically stopped by a brake, which is activated by contact pressure from operation of the detector level.

Dropped cocoon gatherers

The apparatus gathers baskets that have collected all cocoons dropped during end feeding. These are carried to the dropped cocoon separator. These baskets travel intermittently between the cocoon suppliers.

Separator of dropped cocoons

The device accurately distinguishes and separates pupae, dropped middle layer cocoons and thick layer dropped cocoons.

6. Semi-automatic reeling machine

This is a mixed-mode or hybrid machine between the automatic and multi-ends reeling machines. As an application, this type is better for improved reeling efficiency and raw silk quality than the multi-ends reeling machine. The semi automatic reeling machine can be operated with poor quality cocoons, but relies on more labour than the automatic reeling machine.

The cooked cocoons are carried into the groping end part (10) and the end-groped cocoons are removed to the picking end part (11). The correctly picked end cocoons go through the cocoon supplying basket (9) and then to the arranging basin (8). When the size detector of the reeling thread indicates the feeding motion, the picked end cocoons on standby are fed to the reeling thread by a feeding spoon. The reeling thread passes through the jetboute, silk button, first guider, second guider, third guider, fourth guider, size detector, fifth guider and traverse guide, and then it is finally wound onto small reels (4) (Figure 16).

The end dropped cocoons are collected by the cocoon buckets (13) and removed to the cropped cocoon basin, where more reelable cocoons are separated and then poured into the groping end part (10). The principal difference between a semi-automatic reeling machine and an automatic reeling machine is that the cocoon end groping, cocoon end picking and cocoon carrying are manual.

6.3 Methods of silk reeling

a) *Reeling velocity (R.V.)*

Reeling velocity is defined as the winding speed of raw silk on the reel. The reeling velocity is measured as the length of raw silk reeled during 1 minute or as the revolution number of the reel per minute. It is usually calculated by the following formulas.

$$(1)\ \text{R.V. (m/min.)} = \frac{\text{None breaking reelable bave (m)} \times \text{number of end feeding bave (pc/min./reel)}}{\text{Ave. reeling cocoon number per thread}}$$

$$\text{(2) R.V. (m/min.)} = \frac{\text{None breaking reelable bave (m)} \times \text{number of end reeling (pc/min.man)}}{\text{Ave. reeling cocoon number per thread} \times 20 \text{ (reels)}}$$

Where formula (1) applies to the automatic type, formula (2) applies to the multi-ends type.

Number of end feeding :	Automatic type -	1.2-1.5 pc/min./reel
	Multi-ends type -	12-14 pc/min./man

$$\text{Ave. reeling cocoon number per thread} = \frac{\text{required size of raw silk}}{\text{size of cocoon filament}}$$

None breaking reelable bave = length of cocoon filament (m) x cocoon reelability (percent)

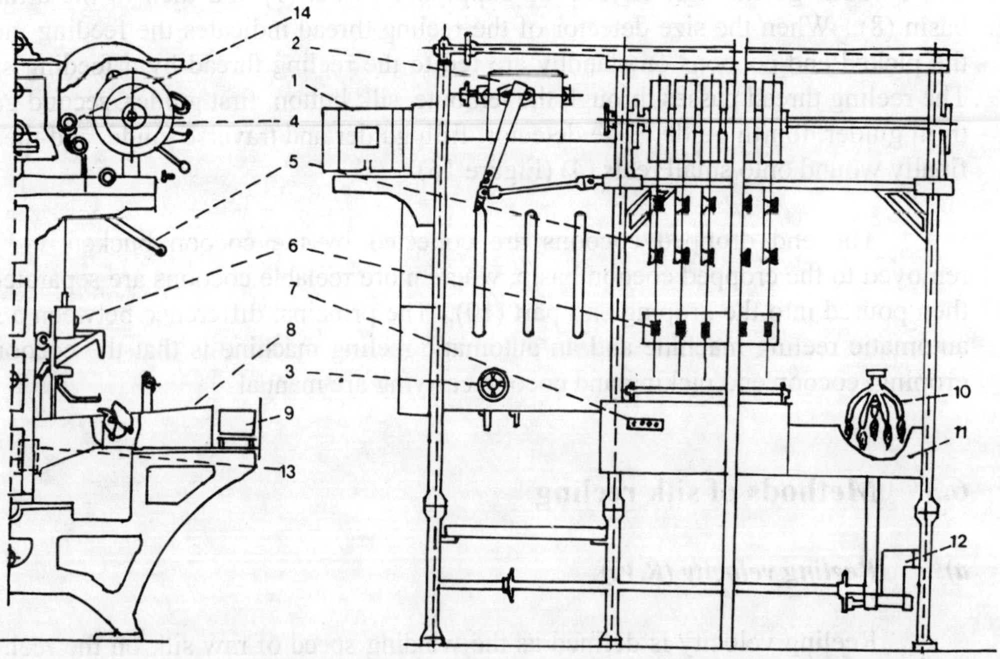

Figure 16. Dimension and structure of semi-automatic reeling machine.

1. Motor, 2. Traverser gear, 3. Chain stepless regulator, 4. Small reel, 5. Stop motion device, 6. Denier control device, 7 Brush, 8. Arranging basin, 9. Cocoon carrier, 10. Brush, 11. Picking bath, 12. Brush drive device, 13. Reconveying bucket (dropped cocoon), 14. Steam pipe.

The velocity of cocoon reeling affects the raw silk yield, reeling efficiency and raw silk quality (see Table 21 and 22). Optimum reeling velocity should be maintained for the best possible product.

b) ***Groping ends***

The high water temperature of the groping ends section dissolves sericin to a great extent and improves the reelability of the cocoon. However, it is apt to reduce raw silk percentage of the cocoon. Alternatively, too low water temperature in this process can reduce the groping end efficiency, the reelability percentage and the reeling efficiency.

Table 21. Effect of various reeling velocities on silk yield and quality with different grades of reelability cocoon (Choe, *et al.* 1971)

Cocoon reelability	Reeling velocity (m/min.)	Raw silk percent of cocoon	Percentage of cocoon reelability	Raw silk quality	
				Neatness (%)	Cleanness (%)
Good Group	90	16.87	72	96.2	96.8
	120	16.80	71	95.2	96.8
	150	16.73	71	93.4	94.5
Fair Group	90	16.36	58	91.9	93.7
	120	16.32	57	90.9	93.8
	150	16.21	55	90.5	93.0
Poor Group	90	16.17	50	91.0	92.3
	120	16.07	49	89.5	90.8
	150	15.70	46	89.0	90.6

Table 22. Relationship between reeling velocity and evenness of raw silk quality (S. Kimura, *et al.* 1964)

Reeling velocity (m/min.)	Average evenness (%)	Low evenness (5)
50	93.5	86.0
80	98.0	82.0
110	88.5	80.2
140	87.3	80.0
170	85.5	78.4

An appropriate temperature range is from 75°C to 80°C, but it will vary. The number of groping end cocoons is about 70-90, but this also varies with the cocoon quality. The pH value of water at this point should be from 6.8 to 7.0. the water should be changed often to prevent discolouration of raw silk.

c) ***Picking ends***

This is a very important function of the reeling operation, which can affect raw silk percentage and reelability. Correct end cocoons should not be permitted in picking ends to minimize cocoon filament waste.

Insufficient picking ends causes the feeding of incorrect end cocoons, breaking down of the reeling thread, and affects the accuracy of the feeding end work.

On the other hand, excessive picking ends reduce raw silk by causing an increase in waste. The picking end with too many cocoons can cause the work to be carried out inaccurately and is apt to increase waste of cocoon filament by repeated work on correct end cocoons. The number of cocoons for a picking end section should be below 40 pcs, and the recommended temperature of the water bath is 40 to 50°C. In automatic end-picking machines, the degree of end picking varies between 80 to 85 percent for good cocoons, 75 to 80 percent for medium ones and 70 to 75 percent for inferior ones.

d) Reeling bath

- **Temperature -** High temperature in the reeling bath increases the breakdown of the reeling thread, cleanness defects and reduces raw silk yield. It improves the reelability of cocoons, the cohesion and the feel of raw silk by reducing the gum in the sericin. The optimum reeling bath temperature depending on cocoon characteristics is about 30 to 45°C.

- **Consistency of water –** Continuous reeling work without changing of reeling water results in water, which is dense, coloured and acid. High density water reduces the cocoon reelability, but increases raw silk percentage making the silk light greyish in colour. Water supplied to the reeling bath should be pH controlled so that the water for each reeling bath may remain the same to ensure colour uniformity. The best consistency is when the pH of the water is about 6.8-7.0, and the water contains a little diluted and floating sericin.

- **Supply of cocoons –** In automatic reeling machines, cocoons are supplied in two ways: rotary type and fixed type. In both types, when the size detector is activated, cocoons are supplied one by one, accurately and quickly. In order to raise the feeding efficiency, it is important to minimize the end-missing cocoon supplier and eliminate the floating cocoons.

- **Maintenance of size detector –** The identification of size by the detector should be accurate. In practice, however, the performance of the detector can be badly affected by improper care, dirt, and abnormal resistance of thread passage and other defects. During operation, the detector should be kept clean and in good repair. Periodic inspections and maintenance should be carried out.

- **End feeding and button** – End-feeding should furnish the end within 1 to 2 seconds without long knots. The hole of the button should correspond to the size of the raw silk.

  The diameter of the buttonhole should preferably be in the following range:

$$39.0\sqrt{S} < D < 52.0\sqrt{S}$$

Where S is the size of raw silk, and D is the diameter of botton hole.

- **Control for breakdown of a reeling thread** – Inefficient processing is considered to be the main cause of frequent breakdown, which in turn decrease reeling efficiency. The reeling velocity has to be controlled so that the total working period of repairing any breakdowns may be less than 60 percent of one pulse operation period. The breakdown ratio by snap reading is about 1.5 percent to under 2 percent.

- **Arrangement of reeling part** - Disordered reeling parts make reeling difficult, reducing the raw silk percentage and the reeling efficiency, while deteriorating the raw silk quality. Reeling conditions of reeled cocoons should be monitored through continuous inspection.

6.4 Quality control during raw silk reeling

The purpose of reeling process is not only to raise the raw silk yield of cocoons and reeling efficiency, but also to improve raw silk quality.

a) **Size development and evenness** - To maintain reeling thread in the required size, the average cocoon number per thread must be adjusted by a check to produce silk thread in the same size throughout all ends during reeling. If the size is different from the required size it should be readjusted by the group size controller. This will reduce size deviation. By improving the accuracy of cocoon supplying each silk thread becomes uniform in size and different skein sizes are reduced. It can also improve the size deviation and evenness of raw silk quality through accuracy of cocoon supplying work and the improvement of cocoon reelability.

b) **Defects** – Defects in raw silk are divided into super major defects, major defects and neatness defects. Defects occur based on reeling conditions. Types of defects and remedies are explained in Table 23.

c) **Cohesion** – Good cohesion raw silk is needed for the warp of silk fabric. Factors that improve the cohesion are temperature, amount of reeling tension, sufficient croissure and good drying of raw silk.

d) **Colour** – The water consistency and temperature of groping end part and reeling part affect the colour of raw silk. Therefore, the temperature control and water supply in the reeling machine should be constantly monitored to obtain a uniform colour of raw silk.

Table 23. Prevention method against various types of cleanness defects

Types of defects	Prevention methods
Waste	1) Control of over or uneven cooking. 2) Prevention of carry over cocoons by remaining in basins during reeling process. 3) Reducing the number of standby cocoons in the reeling basin. 4) Preventing the over swelling of cooked cocoons in the reeling process by controlling the temperature of grope end and picking end bath as well as maintaining a constant temperature.
Large slug	1) To 4) are same as above. 5) Avoid low temperature cocoon drying 6) Controlling the excessive increase of reeling speed and reeling thread.
Bad casts	1) Keep the extra end of cocoon filament short when the correct end picked cocoons are fed to the reeling thread.
Long knots	1) When the broken thread are tied and knotted during the reeling or re-reeling process, the end of the tied and knotted thread must be cut off, leaving only 2 mm.
Cork screws	1) When the reeler has to eliminate a reeled out or bad cocoon during reeling, he should not use force to pull it out but should snip it out with his fingers as quickly as possible.
Split ends	1) Reduce friction effect on silk thread. Pulleys with cracks and splits on the surface should be removed. All pulleys should be perfectly polished to reduce split ends in the reeled yarn.
Long loops	1) Prevention of over or uneven cooking.

CHAPTER 7

RE-REELING AND FINISHING

7.1 Re-reeling

Re-reeling is a process of reeling the raw silk from small reels onto large (standard) reels with a circumference of 150 cm, adjusting the width and weight of skein uniformly. The object of re-reeling is to make standard sized hanks or skeins for marketing purposes. Skeins should have a circumference of 150 cm in accordance with International Standards. Secondly, re-reeling eliminates defects, which might have occurred during reeling.

7.1.1 Re-reeling machine and apparatus

Basically the re-reeling machine consists of various parts and instruments such as the grand reel, reel revolution and stop motion equipment, traverse bar and guider and drying pipe, double thread checker and accessory utensil (see Appendix, Figure 24). The circumference of the grand reel is 1,50 cm and length of the reel rib to be used for five skeins is usually 0.67 meters.

7.1.2 How to re-reel raw silk

1) Pre-treatment

To obtain best results from the re-reeling process, the sericin gummed at the rib edge of the small reel should be swollen and softened through a dipping process.

I. **Infiltration method** by vacuum tank 203 times infiltration with 30-35°C water under 40-60 cm Hg is recommended (see Appendix, Figure 23).

II. **Bath method.** Dipping time normally 10-30 minutes, but should be changed depending upon the temperature and concentration of the chemicals used. Currently emulsified oil is recommended for use as a chemical for re-reeling.

III. **Damping method.** It is sometimes desirable that the hard gummed raw silk on the small reel be covered with a wet cloth.

2) Dipping during the re-reeling process

In the case of a double skein, additional water is occasionally needed when there has been inadequate wetting of the raw silk on the small reel. This is accomplished by use of water spray, dipping and wet cloth, for instance.

3) *Re-reeling process*

There is a tendency that re-reeling velocity is closely related to the number of breaks of the raw silk, re-reeling efficiency and generation of the gummed skein. Optimum velocity recommended for re-reeling is 150-170 rev./min. for 21-denier and 130-160 rev./min. for 14-denier raw silk. The best temperature and relative humidity of re-reeling is around 41°C, 36 percent R.H. Under this condition the moisture content of raw silk skein immediately after re-reeling reaches 6.2 percent and the raw silk skeins are not stuck together (Table 24).

Table 24. Temperature and humidity of re-reeling in the machine (S. Kimura, *et.al.* 1964)

Temperature (°C)	Relative humidity	Moisture content of raw silk skein after re-reeling (%)	Appearance of raw silk skein
31	55	9.7	Gummed
41	36	6.2	Normal
68	29	2.7	Over dried

Raw silk should have about 6-8 percent moisture in it after the re-reeling process is finished. The number of grand reels per labourer should be around 15-20 grand reels for 21-denier and 20 grand reels for 14-denier. This will change if the raw silk size, re-reeling velocity and number of thread breaks are not within expected bounds.

7.1.3 Re-reeling efficiency

The standard of the re-reeling velocity (Re.V.) and the re-reelable quantity of the raw silk can be calculated according to the following formula.

$$\text{Re.V. (rev..Min.)} = \frac{\text{weight of skein} \times 9\,000 \div \text{size of raw silk (d)}}{\text{re-reeling hour} \times \text{workable number of grand reel by a labourer} \times 1.5}$$

For example, 130 x 9 000 ÷ 20.6/8 x 30 x 1.5 = 158 rev./min.

For the calculation of the re-reelable quantity of raw silk by a grand reel for 8 hours, the following formula should be used.

= Size of raw silk (denier) x re - reeling velocity (rev./min.) x 1.5 m
x (0.05 ÷ 450) x 5 skeins x 480 min.
x effective working ratio of re - reeling
= Size of the raw silk x re - reeling velocity x 0.4
x effective working ratio of re - reeling (Normally 85%)
For example, 20.6 x 170 x 0.4 x 0.85 = 1.19 kg.

7.1.4 Silk end tying and skein lacing

The tie end of the raw silk should be tied with cotton year, to find the silk thread end easily during the preparation process for weaving. This is known as silk end tying. Skein lacing keeps the diamond cross originally in the skein place. Five interlacings each at three places on the skein are suitable to prevent tangling (see Appendix, Figure 25).

7.2 Booking and packing

In order to provide proper moisture to the raw silk, the skein should be kept between 20-23°C and 65 to 75 percent relative humidity for several hours prior to booking of the skein.

Standard of skein size: There are two standard skeins, the 70g known as a single skein and the 125-135g size known as a double skein. The double skein is the preferred version.

Booking of the skeins: With the bleached cotton rope doubled with 74 plies of 10s yearn, the 30 skeins must be bundled into a book if they are single skeins (5 line 6 stairs). In the case of double skeins, a book consists of 20 skeins.

Packing of a lot: One bale contains 22-30 books and the standard weight is about 60 kg.

7.3 Storage of silk

Raw silk is vulnerable to heat and sunlight, which may alter its colour, while moisture may cause fungal attack. Thus, precautions must be taken to store raw silk. The storeroom must maintain control over temperature and humidity. BHC powder, chloropicrin, naphthalene and other repellents should be utilized to counter damage from insects. Open skeins should not be stored over long periods; properly packed books and bales can be stored for extended periods under strict atmospheric conditions with no deterioration to the raw silk quality.

CHAPTER 8

UTILIZATION OF BY-PRODUCTS

8.1 Introduction

In addition to the chief products of the sericultural industry, there are a host of waste materials eliminated during manufacturing, which have commercial value During rearing, pierced cocoons and double cocoons are created. Silk reeling generates brushing waste, end-missing cocoons and pupae. Double cocoons are processed to make doupion silk. Other rejected cocoons and waste are re-processed into flow silk and spun silk yarns. The pupae are sold for fertilizer, feed stuff and other agricultural purposes.

8.2 Doupion silk reeling

Natural fancy doupion silk can be made from double cocoons. Current practices also mix inferiors single cocoons with double cocoons in the reeling process (Table 25). Coarse size 225d or 110d doupion silk is reeled by one end only; finer 60d doupion is usually reeled two ends. Reeling parameters are velocity of 230 m/min and the reeling bath temperature of 80-90°C. The reeler gropes the cocoons in the pot with the groping broom held in the right hand and feeds the ends to the thread passing through the guide.

8.3 Reeling of non-mulberry cocoons

a) Tasar cocoon reeling

Tasar cocoons have a compact structure and composition distinct from that of mulberry cocoons. The cooking methods and chemical treatment shown in Table 26 reflect details of Tasar processing. The Sitting type doupion silk reeling machines are used for 110 and 225-denier Tasar silk, and Mult-ends reeling machines or Pedal reeling machines are used for the fine 42-63-denier silk.

The "Natwa" pedal reeling machine used for fine Tasar reeling is made of bamboo and wood. While productivity is low, it is widely used in India because it is simple to operate and requires small investment cost.

The Pedal reeling machine is derived from spinning equipment with twisting and winding of the yarn without a fan system. The machine has 4 spindles and a wooden wheel of 50 cm circumference for winding the yarn. It is driven by a foot pedal and the cocoons are reeled by hand. Preferred sizes of Tasar silk are 40/44d and 60/66d.

Table 25. Reeling efficiency of doupion silk by the sizes of thread (G.K. Kim, *et al.* 1965)

Sizes of thread	Formation of raw material cocoons		Doupion silk (%) of cocoons	Reeling efficiency (gams per person in 8 hours)
	Double cocoons	Inferior cocoons		
225 denier	50%	50%	14.2%	1,006
110 denier	50%	50%	14.3%	659
60 denier	50%	50%	14.0%	459

Table 26. A trial of Tasar cocoon reeling (K.E. Song, 1967)

Cooking		Reeling	Results
First stage	Second stage		
A. Solution (alum 0.2% formalin 0.05% nonionic surfactant 0.03%) Boiling for 30 min.	B. Solution (NaOH 0.1%, H_2O_2 0.2% nonionic surfactant 0.03% Soaking for 24 hrs.	- Sitting type reeling machine (for doupion silk reeling) - Reeling cocoon number per thread : 40 pcs.	- Size of thread (denier) 225 - Weight of reeled silk per reeler : 125.6 gr./1 hr. - Silk ratio of cocoon : 8.25%.

b) Eri cocoon spinning

Eri cocoons are open mouthed with a discontinuous filament, which make them suitable only for spinning. Approximately 90 percent of Eri cocoons are hand spun in Assam, India. The characteristics of Eri cocoon bave are shown in Table 27.

Table 27. A survey of characteristics of Eri cocoon bave (B.H. Choe, *et al.* 1969)

Cocoon		Cocoon bave		Composition of bave	
Weight (g)	: 2.4-2.6	Size (denier)	: 1.77	Fibroin (%)	: 72.2
Shell ratio (%)	: 13	Tenacity (g/d)	: 3.5	Sericin (%)	: 11.9
Length (mm)	: 3.5-5.5	Elongation (%)	: 20.8	Fat (%)	: 1.3
Width (mm)	: 14-25			Moisture (%)	: 14.6

8.4 Manufacture of floss silk

Floss silk may be manufactured from any kind of cocoons, but principally it is processed from pierced, end-missing, and double cocoons. Floss silk is beneficial as paddy against cold weather and as a basis for hand spun yarns. The procedure to create floss silk involves degumming, opening-up and finishing.

8.4.1 Degumming

Cocoons are boiled in hot water with added alkali to remove gum and make them soft. If degumming is insufficient, the opening-up will not be easy: conversely over degumming makes floss silk fragile. An example of chemicals used:

NaOH	:	0.05 percent
$NaHCO_3$	:	0.2 percent
Surfactants	:	0.03 percent
Boiling time	:	70 min.

8.4.2 Opening-up

Degummed cocoons are put in clean water and opened up individually by turning each one inside out to remove pupae and other waste. Several cocoon shells are stacked to form a bog into which the fingers of both hands are placed and spread in the water. These are then hung on a frame. Generally, when a sheet of floss silk reaches around 6 g, it is taken from the frame (see Appendix, Figure 29).

8.4.3 Finishing

When the floss silk is removed from the frame, it is either soaked in clean water or a 1 percent hydrogen peroxide solution for a few hours to refine the colour. The, the liquid is squeezed out and the floss silk is shaped to be air dried in the shade from a pole or a rope.

Table 28. Yield of floss silk and spun silk by different cocoon raw materials (K.E. Song, *et al.* 1975)

Different cocoons	Floss silk % of cocoon	Productivity of floss silk		Hand Spun silk % of floss silk	Spinning efficiency	
		Grams per person in 1 hr.	Index		Grams per person in 1 hr.	Index
Double cocoons	66.3	214	100	97.9	20.0	100
Pierced cocoons	65.7	87	41	99.0	20.4	102
Unreelable, reel-off cocoons	59.7	83	39	96.9	17.6	88
Waste cocoons	46.7	64	30	97.5	18.8	94
Cut cocoons	63.7	120	56	95.5	12.6	63

8.5 Spun silk

8.5.1 Introduction

Short lengths of inferior silk filaments taken from waste material are combed and spun together as silk thread. Spun silk threads are soft but less lustrous, strong and elastic than reeled silks. Spun silk fabric will become fuzzy with wear as the yarn is made from short material.

There are several sources to fabricate spun silk:

1. pierced cocoons, the result of breeding moths that have emerged from their cocoons;
2. double cocoons or doupion, which result when two cocoons have been spun too closely together;
3. floss, brushed from cocoons before reeling;
4. friese, the coarse and uneven silk fibre at the beginning and end of each cocoon;
5. Scrap, the machine waste left over from reeling.

The spun silk manufacturing processes are as follows:

Silk wastes → Degumming → Washing → Dehydration
→ Drying → Beating → Wetting → Opening
→ Filling → Circular Dressing → Spreading → Settling → Drawing →
Combing → Roving → Fine Combing → Drawing
→ Slubbing → Re-spinning → Fine spinning → Doubling
→ Twisting → Gassing → Reeling → Spun Silk

Figure 17. Spun silk manufacturing process

Spun silk is less expensive than reeled silk. Although spun silk has less strength and elasticity than reeled silk because of the shorter staple used, it possesses all the general characteristics of reeled silk. Tub silk fabric, for example, is made of spun silk, yet it gives good service when the quality of the fibre is good. Spun silk is used for shantung, pile fabrics, dress trimmings and linings, elastic webbing, sewing silk, summer weight silks, velvets, umbrella fabrics and insulation.

The waste derived from the processing of spun silk yearn is also used. Such fibre is labelled as waste silk, silk waste or most frequently noil silk and silk noil. Silk noil may be reprocessed into spun yarn ad woven into textured fabrics for draperies, upholstery and sportswear. These fabrics are dull, rough, with a cotton-like appearance and more resilience.

8.5.2 Yarn count of spun silk

The size of spun silk thread is defined in a similar manner to standards used for cotton yarn. For cotton, the term "2/60s" signifies a two-ply yarn consisting of two single strands twisted together, each having a yearn count of 60.

In the case of spun silk the notation has a different meaning. For example, for 60/2 two yearns with a separate yarn count of 120 have been doubled, producing a ply yarn with a new count of 60.

8.6 Wild silk

Silkworms hatched from the moth *Antheraea mulitt,* live on oak leaves instead of mulberry leaves consumed by cultivated species. Oak leaves produce an irregular and coarse filament that is hard to bleach and hard to dye. Wild silk has a tan colour derived from the tannin in the oak leaves. It is commonly woven from this naturally coloured thread and is rarely dyed except in solid shades. Standard wild silk thread is made from eight cocoons and averages 32/34-denier. Wild silk is less lustrous than cultivated silk, since about 11 percent of sericin is removed in the degumming process. Wild silk fabrics are durable, washable and less expensive than pure dyed silk. Typical finished fabrics in this category are rajah, shantung, tussah and pongee.

CHAPTER 9

REELING WATER

Approximately 850-1 000 tonnes of water is used to manufacture 1 ton of raw silk. This chapter covers water issues pertinent to silk production.

9.1 Type of water

All natural water sources contain minerals. The mineral constituents found in greatest abundance are bicarbonates, sulphates, calcium chlorides, magnesium and sodium. Carbon dioxide contents may be greater than expected as a result of the decomposition of organic matter. Three water sources are described here: well/spring water, lake water and surface water/river water.

a) Well water is usually clear with a consistent composition. It may contain sodium bicarbonate, calcium bicarbonate, magnesium bicarbonate, iron and carbon dioxide.

b) Lake water can be clear but may be coloured and slightly acidic. High acidity combined with the presence of dissolved gases makes the source corrosive. Although lake water contains some calcium, magnesium chlorides and sulphates, the concentration of electrolytes is low.

c) Surface water contains sulphates, chlorides, calcium bicarbonate and magnesium bicarbonate. The content of minerals and impurities is determined by the profile of the area from which the water originates, plus seasonal changes. Water quality is affected by the amount of industrial affluent discharged into canals and rivers.

9.2 Impurities

Impurities in water are discussed under the following headings:

a) Turbidity and colour
b) Alkalinity
c) Hardness
d) Iron and manganese

a) ***Turbidity and colour.*** Turbidity may be caused by large or small mineral and organic particles suspended in the water. Mineral particles may include clay, silt, calcium carbonate and silica. Organic particles may include fine vegetable wastes, fats and microorganisms. Sedimentation in tanks or reservoirs is adequate to clear large particles, but filtration is needed for small or colloidal fragments. Filtration using sand or the use of coagulants followed by filtration may be indicated. The colour in water signals the presence of dissolved or colloidally dispersed organic matter of unknown composition augmented colloidal iron or manganese.

b) ***Alkalinity.*** Raw water contains bicarbonates in amounts dependent on their origin. Often, water may contain small amounts of carbonate alkalines. The number of bicarbonate and carbonate ions present may be isolated by titration using phenolphthalein as an indicator and methyl orange as an indicator in the second stage.

c) ***Hardness of water.*** Water hardness is caused by calcium and magnesium salts. Other contributing elements to hardness include metallic ions such as iron and strontium. Note that if metallic ions are present they will be found in minute quantities compared to salts.

Hardness is described as permanent and temporary; where permanent is caused by nitrates, chlorides and sulphates, and bicarbonates, which may be boiled to precipitate into carbonates and removing the hardness, cause temporary.

$$Ca(NCO_3)_2 \rightarrow CaCO_3 + H_2O + CO_2$$

Hardness can be measured when a given amount of potassium palmitate or oleate solution is titrated using a given amount of water, concluding when the shaken liquid produces a permanent lather.

More sophisticated methods are now available where the concentrations of calcium and magnesium ions are quantified by titration with ethylene diaminetetra-acetic acid. Hardness is expressed in degrees, but the actual units of measure vary from country to country.

1 unit ($CaCO_3$ ppm) of United States is equivalent to 0.056°dH of Germany.

d) ***Iron and manganese.*** Iron and manganese may be found in water depending on the source of the contamination. The three types of impurities are: (a) ferrous and manganous bicarbonates in well waters when high amounts of free CO_2 are present, (b) ferrous and manganous sulphates in rivers containing acid mine-waste waters, or (c) iron and manganese mixed with organic waste. Iron is present in some waters at the source, but corrosion in pipelines and storage vessels may taint water.

If a given source is alkaline, iron in the form of ferrous bicarbonate may be removed by aeration when ferric hydroxide is precipitated and carbon dioxide released. Aeration also helps to raise the pH by reducing the content of dissolved carbon dioxide.

$$4Fe(HCO_3)_2 + O_2 + 2H_2O \rightarrow 4Fe(OH)_3 + 8CO_2$$

Manganese can be released in the same way but to ensure rapid oxidation, a pH greater than 10 is required through the addition of lime or caustic soda.

9.3 Methods of water softening

This section looks at industrial water treatment using lime-soda processing, base exchange and demineralization to remove all dissolved salts.

a) ***Lime-soda process.*** In this method, hydrated lime and sodium carbonate are added to precipitate calcium and magnesium ions as compounds of low solubility. For temporary hardness arising from bicarbonates, the reactions are as follows:

$$Ca(HCO_3)_2 + Ca(OH)_2 \rightarrow 2CaCO_3 + 2H_2O$$

$$Mg(HCO_3)_2 + 2Ca(OH)_2 \rightarrow 2CaCO_3 + Mg(OH)_2 + 2H_2O$$

Calcium carbonate has a much lower solubility than magnesium carbonate. When water containing bicarbonates of calcium and magnesium is treated, it is the calcium carbonate, which is first deposited. For permanent hardness, the reaction is:

$$CaSO_4 + Na_2CO_3 \rightarrow CACO_3 + Na_2SO_4$$

$$MgCl_2 + Ca(OH)_2 \rightarrow CaCl_2 + Mg(OH)_2$$

The $CaCl_2$ formed in the last reaction is removed as carbonate by the sodium carbonate. It may be noted that, when temporary hardness is eliminated by this method, no salts remain but permanent hardness, an equivalent quantity of sodium salts is left behind.

b) ***Base or cation exchange.*** This process relies on the exchange of calcium and magnesium ions, which occur when water is passed through a bed composed of complex aluminium silicates commonly known as Zeolites.

If Z represents the silicate, the reaction may be represented as follows:

Temporary hardness $\quad Ca(HCO_3)_2 + Na_2Z \rightarrow CaZ + 2NaHCO_3$

$\quad Mg(HCO_3)_2 + Na_2Z \rightarrow MgZ + 2NaHCO_3$

Permanent hardness $\quad CaSO_4 + Na_2Z \rightarrow CaZ + Na_2SO_4$

$\quad MgSO_4 + Na_2Z \rightarrow MgZ + Na_2SO_4$

In contrast to the lime-soda process, the exchange process replaces the calcium and magnesium ions by an equivalent of sodium even when the hardness is temporary. The anions present are unaffected. The bed is made from Zeolite in granular form, sieved to a particle size around 1-2 mm in diameter; its capacity is limited and will eventually be depleted.

However, the process is reversible and employing a concentrated salt solution can regenerate the bed:

$$CaZ + 2NaCl \rightarrow CaCl_2 + Na_2Z$$

When the calcium and magnesium ions are removed as soluble chlorides, they are replaced by sodium. A cyclic procedure is created of softening, regeneration, water wash (to remove excess salt), softening, etc.

Between the softening and regeneration, it is sometimes desirable to backwash by passing water upwards through the bed for a short time to remove any matter, which has filtered out of the bed. Water from this process may have hardness values as low as 0.5 ppm.

c) ***Demineralization.*** If two exchange processes are used, one for the anions and one for the cations, the content of mineral salts may be reduced to that present in distilled water. Removal of the metal ions may be carried out with one of the sulphonated resins in the "Hydrogen" form, so that water running out contains hydrogen instead of metal ions.

$$CaCl_2 + H_2R \rightarrow CaR + 2HCL$$

$$Ca(HCO_3)_2 + H_2R \rightarrow CaR + 2CO_2 + 2H_2O$$

The resultant carbonic acid breaks down into water ad carbon dioxide, which can be removed by aeration, but the mineral acids remain. Regeneration of the exchange process may be carried out by the use of hydrochloric acid.

Table 29. Ion-exchange types and the quality of water treated (K.E. Song and Y.W. Lee, 1973)

Type of treatment	Original water	Base type	Hydrogen type
pH	6.4	6.8	3.3
M-alkalinity ($CaCO_3$ mg/1)	60	81	0
Acidity ($CaCO_3$ mg/1)	30	0	74
Whole hardness ($CaCO_3$ mg/1)	88	8	2

9.3 Quality standard of water

It is essential to carefully select the quality of reeling water, as it impacts reeling efficiency, raw silk percentage of cocoons and the raw silk quality. Table 31 gives a standard for quality of reeling water and Table 31 shows the pH values and water hardness desirable for cocoon cooking water.

Table 30. Standard quality of reeling water (B.H. Kim, 1983)

Items	Standard concentration	Range of concentration
(1) Colour and cleanness	Colourless and clear	
(2) smell	No smell	
(3) Suspension and sediment	No	
(4) pH of water	6.9	6.6-7.2
(5) pH of water after being boiled	8.3	7.9-8.6
(6) Specific electro conductivity (micrombo/cm)	100	40-300
(7) Hardness (°dH) °dH x 17.85 – $CaCO_3$ ppm	2.0	0.5-4.0
(8) M-alkalinity ($CaCO_3$ ppm)	30	20-40
(9) Total acidity ($CaCO_3$ ppm)	5	3-15
(10) Heavy metal iron (Fe_2O_3 ppm)	Under 0.1	0-0.3
(11) Residue after evaporation (ppm)	85	50-200

Table 31. pH values and water hardness suitable for cooking parts

Parts	Items	Good reelable cocoons	Poor reelable cocoons
Dipping part	pH	4.5-5.5	4.5-5.5
	Water hardness (°dH)	0	0
Low temperature permeating part	pH	5.5-6.5	6.0-7.0
	Water hardness (°dH)	0-2	0-2
Cooking adjusting part	pH	5.5-6.5	6.5-7.5
	Water hardness (°dH)	2-4	1-3
Finishing part	pH	6.5-7.0	7.0-8.0
	Water hardness (°dH)	1-3	1-2

9.4 Analysis of reeling water

a) ***pH.*** A pH meter can measure the value of pH.

b) ***Acidity.*** A 50 ml water sample is taken from the original water and five drops of phenolphthalein indicator added. Titrate with the N/50 of NaCO standard solution until the yellowish colour of the sample water changes to purple.

$$a = b \times \frac{1{,}000}{e}$$

Notice: a, Total acidity ($CaCO_3$)
b, Amount of N/50 NaOH sol. Consumed for titration (ml)
c, Amount of sample water (ml)

* Phenolphthalein indicator: 0.15 g phenolphthalein and 0.05 g thymol blue dissolved in 100 ml of 50 percent ethanol.

c) *M-alkalinity.* A 50 ml of water sample is taken from the original water and 5 drops of methyl red indicator added. Begin titration by using the N/50 N_2SO_4 standard solution. Terminate when the bluish colour of the water sample changes to purple (pH 4.8).

$$a = b \times \frac{1000}{c}$$

Notice: a, M-alkalinity ($CaCO_3$)

b, Amount of N/50 H_2SO_4 sol. Consumed for titration (ml)

c, Amount of sample water (ml)

*Methyl red indicator: 0.02 g methyl red and 0.1 g bromcresol green dissolved in 100 ml water

d) ***Water hardness.*** A 50 ml water sample is poured into a beaker. pH 10 is reached by adding 1 ml of Buffer solution and a few drops of E.B.T. indicator to the water sample, shaking it for a while and then begin the titration by using the E.D.T.A. standard solution. Terminate when the reddish colour of the water sample changes to a bluish colour.

$$a = b \times \frac{1\,000}{c}$$

$$°dH = a \times 0.056$$

|Notice: a, Total hardness ($CACO_3$)

b, Amount of N/50 H_2SO_4 sol. Consumed for titration (ml)

c, Amount of sample water (ml)

* Buffer solution: the total volume to be 1 litre by adding the distilled water after 67.5 g of NH_4C1 is dissolved into 570 ml of NH_4OII.

*E.B.T. indicator: The total volume to be 100 ml by adding the ethanol after mixing 9.5 g of Eriochrom Black T. and 30 ml of triethanolamine.

*E.D.T.A. Standard solution: 3.7225 g of the purified Ethylene diamine Tetraacetic Acid disodium salt is dissolved into distilled water and until a total volume of 1 litre is achieved.

CHAPTER 10

FACTORY PLANNING

10.1 Annual requirement of cocoons

The target of raw silk production in a year should be planned in order to determine the size of the reeling factory. The scale of the factory is also influenced by the properties of raw material cocoons because each has different raw silk yieldability. An exact survey of cocoon quality is required before reaching a decision. Generally, the number and scale of reeling machines for installation is derived from the daily target of raw silk production, which is calculated by dividing the annual target of raw silk production by 300 working days per year.

In this example, the raw silk yield from fresh cocoons is considered for two species. The raw silk percentage of fresh bivoltine cocoons is applied in Japan and the Republic of Korea, while the Renditta of multivoltine ratio of fresh cocoons required for producing 1 kg of raw silk is used in India, Pakistan and Sri Lanka. In the case of 60 000 kg of annual raw silk production, the required amount of fresh cocoons is as follows:

- Multivoltine cocoon – 540 000 kg (Renditta, 9)
- Bivoltine cocoon - 353 000 kg (silk percentage, 17)

As can be seen from the above, the scale of the reeling factory should be differentiated by the raw silk percentage of the cocoon. To summarize, cocoon drying facilities are less useful in tropical countries as fresh cocoon production continues all year round. Temperate Zone countries can only rear silkworms twice or three times a year.

10.2 Equipment for installation

The proper equipment for installation has to be selected by considering the following factors:

1) Properties of cocoons to be provided

Length of cocoon filament, length of unbroken filament, size and weight of single cocoon filament, percentage of eliminated cocoons, percentage of cocoon shell and raw silk percentage.

2) Supply conditions of fresh cocoon

a) Supply times per year of fresh cocoon.
b) Distance, road conditions, vehicle and containers for transportation of fresh cocoons.

c) Quality distribution of fresh cocoons with different crop seasons and production sites.

d) The proper kind and capacity of dryer, adaptable method of cocoon storage, size of cocoon storage room and space for fresh cocoon preparation can be decided based on conditions a to c.

10.3 Location and space

a) ***Location.*** The location of the silk reeling factory should be chosen by evaluating availability of reeling water sources, manpower, electricity and transportation. Ideally, the factory will be sited on a slight hill near a good water supply with adequate drainage, near a densely populated region. The main building of the factory is usually in an east/west direction.

b) ***Building site.*** Space for the building site should be procured to minimize the fixed capital and management costs associated with the idle land, unless future expansion is planned at the same location. Most facilities offer sleeping quarters for two shift operations and refreshment for the workers to increase productivity.

10.4 Quality and quantity of filature water

As water has a major impact on raw silk yield, it is recommended that the factory have access to well water and surface water sources. Quality and quantity of available reeling water is the most important factor in choosing location.

If a site near well water is considered, pumping trials should be conducted over a two-week period during the dry season.

Remember that the amount of reeling water required for production of 1 m/t of raw silk is 1 000 m/t in the Multi-ends reeling machine and 800 m/t in that of the Automatic reeling machine.

10.5 Selection of proper equipment

1) Drying machine

a) ***Chamber dryer or cabinet dryer.*** The delivery and feeding of fresh cocoons to the dryer is carried out manually. Only this type of dryer can execute a small production of fresh cocoons.

b) Hot air dryer

- **Multi-step type**

  The Multi-step type, equipped with an endless chain connected by a wire net is classified into Six, Eight and Ten steps. (More steps need less space for installation.) This type of dryer is vertically divided into three sections: upper, middle and lower each having a different temperature.

This type is suitable for a large-scale reeling factory, which has a drying capacity of 5 000 to 18 000 kg of fresh cocoons per 24 hours.

- **One band type**

This type has one band of cocoon feeding wire net which carries at least one foot of cocoon piling height for drying, while the Multi-step does only 5 cm of cocoon piling height on the cocoon feeding wire net. There is no difference in the drying capacity between both Multi-step and One-band types. The latter needs less space for installation, but it consumes more electrical power for the drying operation.

c) ***Low temperature dryer.*** This dryer has a dual purpose of cocoon drying and storage. A continuous supply of fresh cocoons is needed for completion of one operation. More attention should be given to the operation of the dryer, especially temperature control.

2) **Sorting machine**

A small-scale reeling factory is equipped with a simple type of sorting table. A large-scale factory is equipped with a conveyor or system attached to the sorting machine.

3) **Cooking machine**

The proper capacity of the cooking machine should be selected, since most cooking machines are basically the same in structure. In addition, the devices for aiding the pre-treatment part and cooking control parts should be considered for use, depending upon the quality of cocoons, reelability and cocoon shell percentage.

4) **Reeling machine**

The reeling machine to be selected usually depends upon the quality of the available cocoons.

a) ***Simple reeling machine.*** This type of reeling machine has five to six ends per basin without the Stop motion device. For this machine, the operator is seated and processes poor quality cocoons. The reeling capacity is in the range of 500 to 600 g per basin per 8 hours.

b) ***Multi-ends reeling machine.*** The Multi-ends reeling machine is equipped with 20 small reels per basin. The groping ends, picking ends and cocoon feeding of reeling thread is done manually with the reelers in a standing position. The reeling capacity of this machine is in the range of 800 to 1 000 g per basin per 8 hours. This equipment may be used with all qualities of cocoons.

c) **Auto*matic reeling machine.*** An automatic reeling machine may be used effectively if good quality cocoons with 60 percent reelability, 1 000 m filament length and 20 percent cocoon shell are used. In automatic reeling, the production of raw silk can be increased with less labour, compared to

Simple type or Multi-ends type reeling. Note: Mass production can lead to a reduction in raw silk yield. One set of automatic reeling machines usually consists of 400 reels, but 200 reels to 480 reels can adjust the number of reels per set. The capacity of reeling per set is around 30 kg per 400 reels per 8 hours.

5) Re-reeling machine

The re-reeling machine is composed of large wheels known as "window", which reel five or six skeins per wheel. One window of five skeins wheel has the re-reeling capacity of ten skeins per 8 hours (5 skeins x 2 times = 10 skeins), if one skein weighs around 130 g. It is possible today to find the High-speed re-reeling machines on the market.

6) By-product treatment machine

On an industrial scale, the reeling factories in excess of 1 000 ends require installation of a by-product treatment machine to separate pupae from parchment layer "reeled out cocoons". This function is done manually in small-scale factories.

7) Reel permeation device

The reel permeation device "Vacuuming treatment tank" is necessary for the pre-treatment of small reels for the re-reeling process. This device is especially useful for double skein reels, which have over 70 g of raw silk wound on to a reel.

8) Skein twisting and booking machine

There are two kinds of skein twisting and booking machines. One is the short skein-booking machine by which the skeins are twisted and folded once. Another is the long skein-booking machine by which the skeins are twisted on the turntable and booked without folding.

9) Raw silk testing equipment

The raw silk testing equipment for quality control in the reeling factory is required as follows:

- Seriplane machine
- Seriplane illumination equipment
- Conditioning oven
- Denier scale and Sizing reel
- Rewinding machine

The Serigraph and Cohesion testing machines are not necessary for reeling factories, but are for quality assurance and research institutions.

Table 32. Items of machinery required for a silk reeling factory (Capacity 70 M/T fresh cocoons yearly)

Types of machines	Specifications	Quantity	Capacity
Cocoon drying machine	Hot air blowing and circulating system, three horizontal sections upper, middle and down rooms with automatic electrical thermostats L 12 700 x W 5 300 x H 2,000	1 set	600 kg of fresh cocoons per 8 hrs.
Cocoon cooking machine	Endless chain system. No. of baskets : 70 L 8 550 x W 1 190 x H 1,795	1	200 kg of dried cocoons per 8 hrs.
Cocoon reeling machine	Semi-automatic type with denier detecting system 14 ends per basin L 16 135 x W 1 360 x H 1 795	32 basins	117 kg of dried cocoons per 8 hrs.
Silk re-reeling machine	5 heads per wheel L 18 470 x W 1 360 x H 1 360	40 wheels	45 kg of raw silk per 8 hrs.
Vacuum permeation device	Power : 1.5 hp Φ 750 x H 1 200	1 set	100 kg of raw silk per 8 hrs.
Cocoon assorting machine	Conveyor type	1 set	300 kg of dried cocoons
By-product treating machine	Automatic type L 7 880 x W 545 x H 1 000	1 set	30 kg of dried pellete per 8 hrs.
Silk booking machine	Motor driven	1 set	
Silk skein twisting machine	Hand driven	1 set	
Seriplane winding machine	20 black boards, carrier and Standard Photos included	1 set	
Illumination equipment for inspection	For Cleanness and Neatness test	1 set	
Denier scale	400 revolution and 200 revolution	2 sets	
Sizing reel	Circumference of reel 1.125 m	2 sets	
Rewinding test machine	10 bobbins	1 set	
Boiler		1	2 tons
Oil tank		1	
Water tank		1	
Chimney		1	
Electric equipment		1	
Mechanical tools		1	

Table 33. Building space required for each item of machinery to be installed

Rooms	Space (m^2)	Remarks
Cocoon drying	165	Hot air circulating type : 1 set
Cocoon storage	165	Honey comb tube type and general room type
Cocoon sorting	30	Conveyor type sorting machine : 1 set
Cocoon cooking	60	Auto cocoon cooking machine : 1 set
Reeling	230	Semi-automatic reeling machine : 32 basins
Re-reeling	165	Re-reeling machine "32 wheels
Inspection and finishing	40	Seriplane winding machine, illuminating equipment and rewinding test machine
By-product treating	60	Auto by-product treating machine : 1 set
By-product storage	30	
Boiler	50	Boiler 1 set (2 tons)
Mechanic	30	Tools and spare parts
Raw silk storage	30	
Rest	50	
Dormitory	100	
Officials	50	
Total	1,255 m^2	

CHAPTER 11

RAW SILK TESTING

11.1 Introduction

Silk weaving has reached a very high standard of industrial efficiency. In fact, today a number of varieties of silk fabrics are produced on handlooms and sophisticated power looms. This requires different qualities of raw silk. In order to assist the weaving industry in the selection of the required raw silk, it must be first tested and classified. Further, the raw silk reeling industry requires well-defined standards, which can only be achieved by silk testing. As the demand for silk is global and a number of countries compete in the trade of raw silk, it is necessary that there should be industry standards for raw silk quality so as to enable buyers to purchase raw silk at internationally accepted grades. This is the reason why all raw silk produced should be classified following testing.

The testing of raw silk is based on the procedure laid down by the International Silk Association (I.S.A.). This procedure is quite conservative and the equipment used is consistent with traditional patterns when compared to procedures followed for general textile products. The traditional method has been widely preferred in silk producing and consuming countries of the world. The mechanical testing procedure is similar everywhere, but the compilation of test results and standards for various grades differs slightly from country to country.

In 1914, an essay competition was held in New York for quality silk goods. During 1921-1927, silk classification committees were set up. In 1928 and 1929, International Technical Conferences were held in Yokohama and New York, respectively. All silk producing countries such as Japan, China, Russia, France, Italy, etc., and silk consuming countries such as the United Kingdom and the United States, were represented. The organization of the International Silk Association and the organization of the third International Technical Raw Silk Conference in Zurich in 1949 followed by a fourth one in New York in 1950, contributed to placing silk technology and silk research on a global basis. In 1949-1950, a Bulletin was issued by the International Silk Association (I.S.A.) on International Standard Methods for Raw silk Testing and Classification. In 1961 a revision was made and it was decided to divide them into three categories. A second amendment was made in raw silk testing and classification in 1974. Recently, an electronic testing system was introduced into I.S.A.'s methods.

Two main tests are usually applied, namely the conditioned weight test and the test for quality. The quantity tested from a consignment depends on the total weight. Usually, 5-10 bales of 60 kg each or 20 bales of 30 kg each is the unit of testing used for the export of raw silk.

11.2 Quality test

11.2.1 Visual Inspection

This test is conducted on the whole lot of raw silk, which has to be classified. This is carried out in an inspection room, which is well illuminated. There are three main factors that have to be tested. These are:

11.2.2 Uniformity

In this test, the entire lot is inspected to assess the uniformity of colour, lustre and feel. The result is not recorded in the I.S.A. method, but is classified as good, fair and inferior under the Japanese method.

11.2.3 General finish

Here, considering the presence and degree of a number of defects assesses the general finish of the lot. These defects are:

i. **Re-reeling.** Gummed skeins; gummed spots on skeins; double ends; irregular traverse and partial flack of traverse.

ii. **Finish.** Tangled filament, defective lacing, filament out of place in skein (pulled filaments loose).

iii. **Arrangement.** Lacing of booking card through skeins; non-uniform skeins; wrong twisting; raised filament; streaky filament; cut ends; discoloured skeins; foreign matter on skeins; irregular skeins on book; knots on skeins; skeins or books of different types.

iv. **Damage.** Books of irregular shape; gummed books, soiled filaments; frayed skeins; insect attached skeins; musty skeins, etc.

The results are expressed in terms such as good, fair, poor or inferior. Outstanding defects are mentioned in the Test Certificate.

11.2.4 Nature

The degrees of Colour and Lustre and the Smoothness or Hand of the lot are inspected and indicated in the following manner:

i. Colour: light, medium or deep
ii. Lustre : bright, medium or dull
iii. Hand : smooth, medium or rough

11.3 Sample test

After completion of the test on the whole lot, the following tests have to be conducted. To conduct these tests, it is necessary to take out 50 skeins off the lot if the skeins weigh below 120 g. Alternatively, 25 skeins are taken if the skeins weight over 120 g. Having extracted the required number of skeins, the following test is carried out:

11.3.1 Winding test

Equipment required:

- Winding frame
- Wheels and bobbins

Number of sample skeins drawn out for this test should be 40 skeins out of 50 skeins for a lot which weighs approximately 70 g each and 20 skeins out of 25 skeins drawn out for each skein weighing approximately 140 g.

When winding is started only the top half of the sample skeins should be wound. The winding should be carried out at a predetermined speed for a specific duration. The number of breaks that occur should be counted and noted. When breaks occur, it is necessary to note the cause of each break and this should be recorded.

Table 34. Average speed and winding period for winding test

Size under test	Preliminary winding	Average speed (metre/min.)	Winding period (minutes)	
			70 g skeins	140 g skeins
12 denier or finer	10	110	60	120
13-18	10	140	60	120
19-33	10	165	60	120
34-69	5	165	30	60
70 or coarse	5	165	20	40

The winding speed and duration of winding has also to be adjusted according to the denier of the raw silk being tested. For reference purposes, the different time and speed and other requirements are shown in Table 34.

11.3.2 Size deviation test

Equipment required:

- Sizing reel (1:125 metres in circumference)
- Balance
- Denier scale

Only a fixed length from each skein is taken for the testing. The highest degree of deviation is noted and compared with the average size deviation of the conditioned weight. When testing raw silk of 33-denier or finer in size, 200 skeins of 450 metres each are taken out by picking out four skeins out of each of the 50 sampling skeins taken from the testing lot. In the case of 34-denier and over, 400 sizing skeins are extracted by picking eight skeins of 112.5 metres in length each from the 50 sample skeins (total length 45 000 metres).

Test

The entire lot of sizing skeins should be in ten separate lots.

Each sizing skein should be weighed on a quadrant scale in each lot separately. The total weight of all the skeins in each lot is therefore obtained. Table 35 indicates the size deviation of the skeins being tested. This table covers the number of skeins in a lot, the graduation in the scale for weighing each skein and all the skeins of a lot together.

Table 35. Number of skins in a lot and graduation in scale

Sizes	No. of skeins in a group	Accuracy of scale		Permissible range of denier
		For 1 skein	For group	
330 denier or finer	20	0.5 denier	0.5 denier	1.5 denier
33-49 denier	40	1.0 denier	2.0 denier	4.0 denier
50-99 denier	40	2.0 denier	2.0 denier	8.0 denier
100 denier or coarser	40	5.0 denier	2.0 denier	19.0 denier

If there is a difference in weight beyond the permissible denier indicted in the above table when the weight of the individual skeins are totalled and compared with the result obtained by weighing all the skeins together, the weighing operations both of single skein and the whole lot must then be repeated. The number of individual sizing skeins of each denier size should be noted in the table of frequency distribution.

$$S = \sqrt{\frac{\sum f_i (t_i - M)_2}{N}}$$

The size deviation can be obtained by the following formula:

Where $t_1, t_2, t_3 \cdots\cdots$, t_n are the sizes of the sizing skeins in a denier; $f_1, f_2, f_3 \cdots\cdots$, f_n are their frequencies.

N is the total number of sizing skeins.
M is the calculated average size.

The results of the size deviation test are calculated down to two decimal places.

- Maximum deviation test : the difference between average size and average of four thickest sizes and also the difference between average size and average of four thinnest sizes are determined. Comparison is then made of both differences and the result of maximum deviation determined as the larger one. The result of maximum deviation is calculated down to one decimal place.

- Average size test : All the sizing skeins are then placed in the conditioning oven and dried at 140°C. Until it reaches a constant, weighing is undertaken after the first ten minutes of drying.

Table 36. An example for size deviation test

Denier	Frequency of each 20 skeins group 1	2	3	4	5	f	Deviation (x)	fx	fx^2
16-1/2			1		1	2	-10	-20	200
	11					2	-9	-18	162
17-1/2	1	1			1	3	-8	-24	192
		11		11	1	5	-7	-35	245
18-1/2	1		11	11	1	6	-6	-36	216
	1	1	1	111	11	8	-5	-40	200
19-1/2		1	11	1	1	4	-4	-16	64
	1	11	1	1	11	7	-3	-21	64
20-1/2	11	111	11	1	1	9	-2	-18	36
	1	11	1	1111	11	10	-1	-10	10
21-1/2	11111	111	1111	1111	1	17	0	0	0
	111	111	1		11	9	1	9	9
22-1/2			1	1	1	3	2	6	12
	1		11	1	1	5	3	15	45
23-1/2		11		1	11	5	4	20	80
	1			1		2	5	10	50
23-1/2	1				1	2	6	12	72
				1		1	7	7	49

Group	Sum of 20 skeins	Denier wt. of group
1	406.0D	404.5D
2	403.0	404.5
3	409.0	410.0
4	399.5	400.0
5	403.0	402.0
sum	2,020.5	2,021.0
mean	20.205	20.21

Sum of fine deviations	-238
Sum of coarse deviations	+79
Σfx	-159
$(\Sigma fx)^2$	25281
Σfx^2	1706
Middle of deniers	21D
Deviation of class	0.5D
No. of total skeins	100pcs

$$\text{Size deviation (denier\_)} = 0.5\sqrt{\frac{1}{100}(1706 - \frac{25281}{199})} = 1.90$$

Maximum size deviation (denier) =

$$\text{o Ave. of 4 coarsest size} = \frac{24.5 + 24.0 + 24.0 + 23.5}{4} = 24.0$$

$$\text{o Ave. of 4 finest size} = \frac{(16.0 + 16.0 + 16.5 + 16.5)}{4} = 16.25$$

$$\text{o Coarsest deviation} = 24.0 - 20.205\ (\text{Ave. Size}) = 3.795$$

$$\text{o Finest deviation} = 20.205\ (\text{Ave. Size}) - 16.25 = 3.945$$

Therefore, the maximum size deviation is 3.945.

11.3.3 Evenness test

The test is carried out with test samples of a fixed length using a Seriplane. These test samples represent fine passages and coarse ones, divided into 3 groups (Evenness Variation, **I, II** and **III**) according to the degree and fequency of size variations.

- ° Evenness Variation **I:** The intensity of variation greater than the V_0 panel but does not exceed V_1 panel of the Standard Variation Photographs.
- ° Evenness Variation **II:** The intensity of variation greater than the V_1 panel but does not exceed the V_2 panel of the Standard Variation Photographs.
- ° Evenness Variation **III:** The intensity of variation, which includes all the variations greater than the V_2 panel of the Standard Variation Photographs.

Panel - A panel is a section of raw silk 127 mm wide by 457 mm long uniformly wound from a bobbin on to an inspection board.

In the Japanese method, Variation I is applied not for the major test or auxiliary test but for the optional text.

°Apparatus and equipment: Seriplane (127 x 457 mm), Standard Photographs and Illumination room.

° **Sample:** The sample for the test consists of a total of one hundred panels from 50 test samples taken at the rate of two panels from each test sample.

The thread is spaced on the inspection panel according to the size under test as follows:

9 denier or finer	..........................133 threads per 25.4 mm
10 to 12 denier	..........................114 " "
13 to 16 denier	..........................100 " "
17 to 26 denier	.... 80 " "
27 to 36 denier	.... 66 " "
37 to 48 denier	.... 57 " "
49 to 68 denier	.... 50 " "
69 to 104 denier	.... 40 " "
105 to 149 denier	.... 33 " "
150 to 197 denier	.... 28 " "
198 denier or coarser	.... 25 " "

Test:

The test is conducted by the estimator from a position of about 2 metre distance directly in front of the inspection panels, which are placed in such a way that the panel receives the same intensity of light by indirect lighting, over the entire board. On any one side of the inspection board, each stripe found on each panel is carefully compared with the standard variation photographs and the intensity of variation is determined. The frequency of evenness Variation **I, II** and **III** found with the 100 panels is recorded separately. The record shows each total frequency of variations grouped in Evenness Variation **I, II** and **III.**

The evaluation of evenness in the Indian method (Table 41) is indicated by percentage. Percentage is indicated to the nearest 5 percent starting from 100 percent to 50 percent.

Below 50 percent, it is made to the nearest 10 percent. The record then shows the estimated evenness percentage of each panel, the average evenness percentage of a total of one hundred panels and the low evenness percentage of low panels corresponding to one-quarter of the total panels inspected.

11.3.4 Cleanness test

This test is conducted to ascertain Super Major Defects, Major Defects and Minor Defects. Each defect carries penalty points and the difference of the total penalty points from 100 gives the test result.

° **Definitions**

Cleanness Defects: These are categorized into three general groups, viz., Super Major Defects, Major Defects and Minor Defects.

Super Major Defects: All major defects in length or size which are ten times larger than the minimum size of Major Defects are named Super Major Defects.

Major Defects: These are divided into five as follows:

i) **Waste.** This is a mass of tangled cocoon filaments or fibres attached to the yarn.

ii) **Large slugs**. These are somewhat thickened places in the thread 7 mm and above in length, or very badly thickened places shorter than 7 mm.

iii) **Bad casts**. These appear as abruptly thickened places in the yearn due to the cocoon filaments not properly adhering to the raw silk yearn, or caused by feeding more than one cocoon filament at a time.

iv) **Very long knots.** These are knots, which have loose ends, 10 mm and over, or those made by incorrect tying of threads.

v) **Heavy corkscrews** are places in which one or more cocoon filaments are longer than the rest, and give the appearance of a very coarse and large spiral.

Minor Defects: The minor defects are subdivided into four as follows:

(a) **Small slugs**, which are considerably thickened places in the thread from 2 to less than 7 mm in length, or extremely thickened places less than 2 mm in length.

(b) **Long knots** are knots, which have loose ends below 10 and more than 3 mm in length.

(c) **Corkscrews** are places in which one or more cocoon filaments are longer than the remainder, and give the appearance of a thick spiral.

(d) **Long loops** or loose ends are loops or split ends, 10 mm and above in length, when measured along the filament.

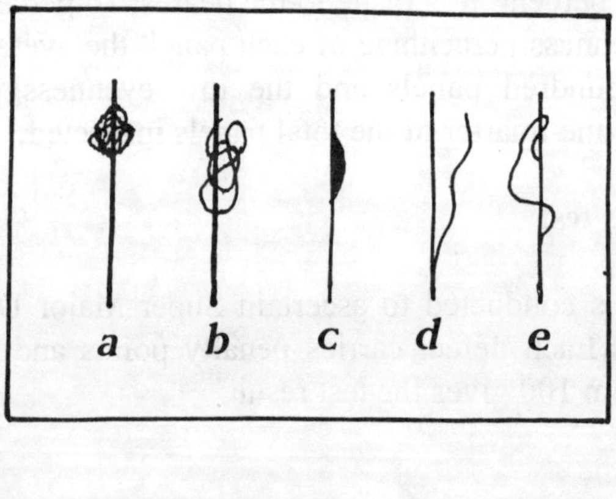

Figure 18. Various cleanness
a) Waste, b) Slug, c) Bad casting, d) Split ends, e) Large loop

Apparatus and equipment: Standard Photograph for cleanness, Seriplane and lighting equipment.

Sample: The same as given under Evenness test.

Test

In this test the inspector stands in a position of about 0.5 metres (2 feet) directly in front of the inspection panels, under the same conditions of lighting as for the cleanness test. The actual number of cleanness defects of each class and kind of defect described above, are counted on the yarns on both sides of the inspection panel, omitting the parts on its edges. The class and kind to which each defect belongs is determined by comparing it with the Standard Photographs for cleanness defects. The record should indicate the number of defects in 100 panels found by testing and also show the cleanness percentage, which is determined by deducting from 100 percent the total penalty calculated by penalizing each defect with the following rate:

For each super major defect .... 1.0 percent
For each major defect .... 0.4 percent
For each minor defect .... 0.1 percent

11.3.5 Neatness test

Neatness defects : Imperfection in raw silk yarn, which are smaller than those described as minor cleanness defects are known as neatness defects. Nibs are small thickened places or spots in the yarn less than 2 mm in length. Loops are small open places in the yarn caused by the excessive length of one or more cocoon filaments, less than 10 mm in length when measured along the filament. Hairiness and fuzziness are the conditions of yarn, which show small loose ends of less than 10 mm and fine particles of cocoon filaments protruding from the yarn. Small knots are knots, which have loose ends, less than 3 mm in length. Fine corkscrews are places in which one or more cocoon filaments are longer than the remainder and give the appearance of a spiral.

Apparatus and equipment: The Standard Photographs for neatness defects, Seriplane and lighting equipment.

Sample: The same as given in the Cleanness test.

Test

The test is conducted by the inspector, from a position of about 0.5 metres (2 feet) distance directly in front of the inspection panels, under the same conditions of lighting as for the neatness test. Each panel on any one side of the inspection board is carefully compared with the Standard Photographs for neatness defects and its neatness value is estimated in percentages. From 100 to 50 percent, the estimate should be to the nearest 5 percent. Below 50 percent, it should be made to the nearest 10 percent. The record should indicate the estimated neatness percentage of each panel, the average neatness percentage of a total of one hundred panels and the low neatness percentage represented by the average percentage of the low panels, i.e. on fifth (20 panels) of all panels examined. In the Japanese method, Cleanness and Neatness is represented as a mixed item by deducting the rate of Neatness results from the Cleanness percentage as shown in Table 37.

Table 37. Deducting points by Neatness results

Neatness	Deducting points	Neatness	Deducting points
Above 80	0		
75	0.25	50	1.5
70	0.5	40	2.0
65	0.75	30	2.5
60	1.0	20	3.0
55	1.25	10	3.5

11.3.6 Tenacity and elongation test

To test the strength of the raw silk, the breaking point (g per denier) and the degree of elongation (percentage) is carried out on the Serigraph. This test is conducted in a room, which is maintained at a standard temperature of 20ºC and a humidity of 65 percent.

Apparatus: Serigraph, sizing reel and scale.

Sample: Ten test pieces taken out of 50 test pieces. From each one of these ten test pieces, ten sizing skeins shall be prepared for tenacity and elongation test.

Test

The sizing skeins to be tested are placed in a room, where standard humidity can be maintained, for a sufficient amount of time to allow them to become adjusted to standard conditions.

Each sizing skein is then tested for tenacity and elongation using the serigraph, which is placed in the room under the same standards of humidity. The tenacity is expressed in grams per denier, while the elongation is expressed in percentage of total stretch of the portion tested. The result is indicated by the average results of ten sizing skeins. The result of tenacity is calculated by omitting the figures after the second decimal.

11.3.7 Cohesion test

By means of the Duplan cohesion tester, the number of frictions required to split silk thread for the purpose of examining the state of cocoon filaments sticking together, can be counted. This test is conducted in a room kept at standard temperature and humidity.

Apparatus: Duplan cohesion tester.

Sample: the sample for the test should consist of 20 test pieces taken out of 50 test pieces. The yarn should be free from any cleanness defect or apparent evenness defect in the portion, which is to be tested.

Test

The test is performed in a room where standard humidity can be maintained during the test. The maximum speed of stroke should be 140 strokes per min. The machine should be stopped after every ten strokes and every single yarn inspected very carefully to see if there are any open places. As soon as ten different open spaces, 6 mm long and above are observed, they are recorded against the number of strokes which should be considered as the strokes of the thread opened. The record of the test is the average number of strokes of 20 test pieces. The result of cohesion is recorded by omitting the decimal.

11.4 Quantitative test

11.4.1 Conditioned weight test

When weighing the gross weight and the dried weight of raw silk, the conditioned weight is calculated by adding 11 percent (International Standard Regain) to the dried weight.

The conditioned weight of raw silk is obtained as follows:

All skeins of raw silk for the test are put into a room with a constant temperature and humidity for 12 hours. 20 skeins are picked up for a test unit and are then weighed separately as the original weight.

Two skeins are selected from the sample for the conditioned weight test.

The sample skeins are dried in an oven one by one with the drying temperature at 140°C. The drying is continued until the different at every weighing (every 5 minutes) is within 0.1 g.

Moisture regain is calculated as follows:

If the different between both skeins in moisture regain is over 0.5%, another one is tested again.

$$\text{Moisture regain (\%)} = \frac{\text{(original weight - dry weight)}}{\text{dry weight (g)}} \text{x} \, 100$$

The total conditioned weight of the unit could be obtained from the total original weight multiplying by the following coefficient.

The coefficient of the conditioned weight is obtained by:

$$\text{Coefficient} = \frac{1.11}{1 + \text{moisture regain (\%)}}$$

Notice:

$$\text{Moisture content (M.C.)} = \frac{W - W'}{W} \times 100$$

$$\text{Moisture regain (M.R.)} = \frac{W - W'}{W'} \times 100$$

Where, W is the original weight and W' is the dry weight.

11.5 Raw silk classification

11.5.1 Grades

For classification purposes, raw silk is divided into three categories according to their sizes:

1st category	:	18 denier and below
2nd category	:	19 to 33 denier
3rd category	:	34 denier and above

The grades are expressed in the following order in all categories of raw silk:

4A, 3A, 2A, A, B, in I.S.A. Standard (Table 38-1, 2,3)
and 5A, 4A, 3A, 2A, A, B, C, D in Japanese method (Table 40-1, 2,3)

11.5.2 Grading with the major tests

The grade of a lot in sizes 33 denier and finer (categories I and II) are determined according to thc lowest respective percentage of its Size Deviation, Evenness I, Evenness II, Cleanness, Average Neatness and Low Neatness. Whereas the grad eof a lot in sizes 34 denier and coarser (category III) are determined according to the lowest respective percentage of its Size Deviation, Maximum Size Deviation, Evenness I, Evenness II, Cleanness, Average Neatness and Low Neatness as tabulated in the classifications table for each category.

Should any one or more of these results fall below the minimum limits prescribed for a grade, the lot is relegated to the lowest grade.

11.5.3 Degrading with the auxiliary tests

a) If any one of the Maximum Size Deviation, Evenness III, Winding, Tenacity, elongation or Cohesion tests of a lot in sizes 33 denier and finer, or if any one of the Evenness III, Winding, Tenacity or Elongation tests of a lot in size 34 denier or coarser, is found to be lower than the required value listed in the class of auxiliary tests, then the grade provisionally established in accordance with the preceding paragraph is lowered by as many grades as the difference that exists between the required auxiliary test class and the class actually found. Any difference that is more than one class is deemed as one class is deemed as one class difference with respect to the Maximum Deviation and Evenness Variation III of a lot in sizes 33 denier and finer and Evenness Variation III of a lot in sizes 34 denier and coarser. In case two or more auxiliary test classes are found to be lower than the corresponding values listed in the classes of the auxiliary tests, then the lot is declassed to the lowest auxiliary test class actually found.

b) In case the result of the visual inspection of a lot is found to be "Slightly Inferior" in its general finish and/or the result of the skein finish inspection in the Winding test of a lot is found to be "Poor", the grade of the lot is the one below that as has been determined in accordance with the preceding paragraphs.

c) In case the result of the visual inspection of a lot is found to be "Inferior" in its general finish, or in case the number of breaks in the Winding test exceed the limits mentioned below, the lot is determined as B grade (D grade in the Japanese method).

12 denier or finer	....	breaks 50(75)
13 to 18 denier	....	breaks 40(65)
19 to 25 denier	....	breaks (60)
19 to 33 denier	....	breaks 35
26 to 33 denier	....	breaks (55)
34 to 69 denier	....	breaks 25(45)
70 denier or coarser	....	breaks 20(35)

11.6 Miscellaneous tests

11.6.1 Boil-off test for raw silk

a) <u>United States Method</u>

° **Object** – The purpose of the Boil-off test on raw silk is to determine the percentage of sericin and water-soluble substance which the silk contains.

° **Apparatus –**

(a) Boil-off kettle – A suitable receptacle for boiling off the silk.

(b) Oven – Conditioning oven with forced ventilation efficient to dry the skeins within the time specified, positive value control, capable of drying the sample skeins at 140°C. The conditioning ovens should be equipped with a balance arranged to weigh the skeins with an accuracy of one centigram while suspended within the drying chamber, the hold of the skeins to be of such type as to insure free access of the dry air to all skins.

° **Water** – The water used for all parts of the test should be zero hardness.

° **Soap -** The soap should be properly saponified "neutral" soap in chip or flake form.

Chemical requirements

Moisture and volatile matter at 105°C, max. percent	....	10.0
Sum of free alkali or free acid, total insoluble matter in alcohol, and sodium chloride, max. percent	....	4.0
Free acid calculated as oleic acid, max. percent	....	0.2
Free alkali, calculated as NaOH, max. percent	....	0.2
Insoluble matter in water, max. percent	....	1.0
Titer of mixed fatty acids prepared from soap	....	16 to 26°C
Anhydrous soap content, min. percent	....	85.0

° **Samples -** Ten skeins, not more than one skein from one book, are taken to represent a five or ten bale lot. Approximately 10 g are removed from each skein. The silk removed from the original skeins is grouped into two parts of approximately 50 g each and marked, "part number one and part number two".

° **Dry weight before boiling** - Dry samples Part 1 and Part 2 separately in a conditioning oven at 140°C to constant weight as determined by successive weighings at five minute intervals. The first weighing is made at the expiration of the first fifteen (15) minutes of drying. The second weighing, made five minutes after the first weighing, is taken as the dry weight provided that the loss between the successive weighings does not exceed one centigram.

° **Boiling** - The samples are boiled for 45 minutes in a one percent soap solution. The weight of the soap should be 25 percent of the weight of the sample and the water 100 times the weight of the soap.

° **Rinsing** = Rinsing should take place in two baths containing water at 60°C, the volume of which should not be less than 25 times the weight of the sample.

° **Boiling – Second boil** - The samples are boiled again for 30 minutes in 2.5 litres of water per 100 grams of silk, i.e. 25 times the original weight of the sample (no soap).

° **Rinsing** – Repeat the rinsing operation.

° **Centrifuge and dry at room temperature.**

° **Dry weight after boiling** – Dry and weigh the samples using the same procedure as described above.

° **Loss in boiling-off** – The difference between the dry weight of the combined weight (Parts 1 and 2) before boiling and the dry weight of the combined weight (Parts 1 and 2) after boiling is the loss in boiling-off. The percentage of loss in boiling-off should be calculated individually per Parts 1 and 2. If the percentage of loss for each of the two parts differs by more than 1 percent, the test should be repeated. The average results of the first test and the repeat test should be reported.

b) European Method

° **Object** – The object of the boil-off test on raw silk is to determine the percentage of sericin and water-soluble substances, which the silk contains.

° **Apparatus** –

(a) Boil-off kettle – A suitable receptacle for boiling off the silk.

(b) Oven – Conditioning even with forced ventilation, positive valve control, capable of drying the sample skeins at 140°C. The conditioning ovens should be equipped with a balance arranged to weigh the skeins with an accuracy of one centigram while suspended within the drying chamber, the holder of the skeins to be of such a type as to insure free access of the dry air to all skeins.

° **Soap** – The soap used should be a neutral soap made from pure olive oil or olive oil foots and soda.

° **Samples** –

(a) Ten skeins, not more than one skein from any one book, are taken to represent a five or ten bale lot. Approximately 10 g is removed from each skein. The silk removed from the original skeins is grouped into two parts of approximately 50 g each and marked, part number 1 and part number 2.

(b) In the case of a single bale, five skeins, not more than one skein per book, should be taken. Two samples of approximately 10 g each are removed from each skein. These ten samples are grouped into two parts, as in (a) above.

° **Dry weight before boiling** – Dry samples part 1 and part 2 separately in the conditioning oven at 140°C to a constant weight as determined by successive weighing at five minute intervals. The first weighing is made after 15 minutes. Weight is considered as constant when the difference between two consecutive weighings is less than or equal to 5 centigrams. The final weighing is taken as the dry weight.

° **First boil** – Boil parts 1 and 2 for 30 minutes in a soap solution composed as follows:

Weight of soap	:	25 percent of net weight of silk.
Concentration	:	8 g soap per litre of water which should be demineralized, or at least at 0° (hydrotimetric)
Volume of solution	:	35 times the weight of silk.
Carbonate sodium (CO_3Na_2)	:	0.3 g per litre of water in the solution.

° **Rinsing** – Rinse by stirring and squeezing the silk by hand in demineralized water (or water at 0° Hydrotimetric). The temperature should be 60°C and the volume of water at least 25 times the weight of the silk.

° **Second boil** – Boil parts 1 and 2 a second time for 30 minutes in a solution having the same composition as the first one, except that no carbonate of soda is added.

° **Rinsing** – Rinse the silk in each of two baths, each time stirring and squeezing by hand.

First bath: To consist of demineralized water (or water at 0° Hydrotimetric) at 60°C with addition of 2 or 3 g of CO_3Na_2 per litre. The volume of water must be at least five times the weight of the silk.

Second bath: To consist of water at 60°C without carbonate of soda.

° **Drying** – Centrifuge and dry parts 1 and 2 for 2 hours at room temperature.

° **Dry weight after boiling** – Dry and weigh the samples using the same procedure as described in the above paragraph.

° **Loss in boiling-off** – The difference between the combined dry weights of parts 1 and 2 before and after the boilings expressed as a percentage of the combined dry weights as found in the above paragraph is the loss in boiling-off.

11.7 Exfoliation test for raw silk

° **Definition** – Exfoliation in raw silk is the undesirable property of the individual filaments of silk split into very fine fibrils. These fibrils initially do not absorb dye as readily as the main fibres and show up after dyeing as fine white fuzz on the surface of the silk yarn and fabric. The Exfoliation Test, as outlined below, is made on raw silk and the degree to which the raw silk tends to exfoliate can thereby be determined.

° **Sample** – The sample for the Exfoliation Tests consists of twenty original skeins drawn at random, an equal number from each bale of the lot, from different parts of the bale. Not more than one skein from any one book except in cases where the lot is composed of fewer than twenty books, and then more than one skein may be drawn from one book as is necessary to make up the twenty skein sample. The sample skeins are wound on regular winding bobbins for a period of twenty minutes.

° **Preparation of panels** – Twenty panels, one from each sample bobbin, is reeled on the metal frames on a specially modified seriplane equipped for this purpose. The size of the panels is 4 and three-quarters inches high and 3 and one-quarter inches wide. The threads are spaced at 25 threads per inch. Each frame is made to hold five panels, making a total of four frames of five panels each for each test. The specially modified seriplane is so constructed that the silk may be reeled on either one or on two frames at the same time. After the silk is reeled on the frames, the frames are placed in special racks for degumming and dyeing. The degumming and dyeing racks are made to hold eight frames or two lots each. Either four or eight frames (one or two lots) may be degummed and dyed at one time.

° **Degumming and dyeing** – The degumming and dyeing are done at the same time in a solution made up as follows:

> Approximately twenty-four litres of water are put into the degumming and dyeing tank and the water is heated. Sixty g of sodium metasilicate are thoroughly dissolved in one litre of hot water and poured into the tank. Thirty g of dye are thoroughly dissolved in another litre of hot water and poured into the tank. The solution in the tank is then stirred thoroughly. A special dye known as United States Testing Company Sky blue S due is used. The thermometer should be left suspended in the solution at all times. When the temperature of the solution has risen to 195°-200°F, the racks holding the frames are placed in the solution and left there for 20 minutes. During the 20minute period, the temperature of the solution should be maintained between 195°-200°F. The same solution may be used a second time for one or two additional lots (four or eight frames). If the same bath is used a second time, tem g of due are added to the solution, but no additional sodium metasilicate is required. The added dye must be dissolved thoroughly in one-half (½) litre of water before being poured into the solution. Sufficient hot water should be added to the solution to maintain the level of the solution high enough to cover the frames. The same solution should not be used for more than two degumming and dyeing operations. It is necessary that the tank be drained and thoroughly rinsed at the end of every day even though the solution has been used for only one degumming and dyeing operation. After 20 minutes, the rack is removed from the degumming and dyeing solution and the frames are rinsed in the rinsing tank to remove excess dye. The rinsing tank is filled with warm water and rinsing is carried out by dipping the rack containing the frames up and down in the warm water three or four times. The silk on the frames is then dried thoroughly. The silk can be dried at room temperature or exposing the silk on the frames to a warm air blast or dried on the frames in a suitable warm oven. If the silk is dried in an oven, care must be taken not to scorch the silk.

° **Preparation of the panels for inspection** – It will be found that the silk threads on the frames will have a tendency to cling together in groups of three to six threads after the degumming and dyeing operation. After the silk has dried, the threads must be carefully separated before the silk is inspected. The separation of the threads can best be done while the frames are on the inspection racks. The threads can usually be separated by running a smooth glass rod, or similar instrument, horizontally across the threads near the top and bottom of the frame. In some cases, it may be necessary to separate some of the threads by means of a needle run vertically between the threads. These operations must be done on both the front and back surfaces of the frame. Extreme care must be taken in separating the threads in order to prevent breaking the filaments or threads. A limited amount of experience and practice will show the best way to separate the threads.

° **Inspection and grading of the panels** – The inspection and grading of the panels is done on the American Standard Inspection Rack. The two special holders are suspended over the top of the overhead light reflector so that the exfoliation frames can be placed in front of the light reflector, the top of the frame being approximately 1 and five-eighths inches below the forward edge of the reflector. A black seriplane board is placed in the upper brackets of the seriplane inspection rack behind the exfoliation frames. The seriplane board is inclined towards the inspection frames to an angle of approximately 40° from the vertical. The angle may be varied slightly so that the silk thread defects can be more easily observed. The exfoliation frames are maintained in a vertical position. Each of the twenty panels is given a rating by comparison with the standard photographs rated as follows: 90-80-70-60-50-30. Each panel is compared with the standard photographs and is given one of the following ratings in accordance with the standard: 100-95-90-85-80-75-70-65-60-50-40-30-20-10. If any filaments have been broken in the separation of the threads which have clung together after degumming and dyeing, care should be taken that the broken filaments are not mistaken for exfoliation and penalized. This difference can be detected after some experience in inspecting panels.

° **Calculation and rating reports** – The general average is obtained by calculating the average of the twenty individual ratings of the twenty panels. The penalty average is obtained by calculating the average of the lowest five panels.

° **The classification degree** is the average of the general average and the penalty average. The rating is based on the classification degree as follows:

Perfect	95 and higher
Excellent	85 to 94.99
Good	75 to 84.99
Fair	65 to 74.99
Poor	50 to 64.99
Very poor	10 to 49.99

Dye for exfoliation test, United States method

United States Testing Co., "Sky Blue S. Dye". This dye has the following formula:

	Class	Colour Index
Alphazurine A : 16 percent	acid	714(NAC)
Wool Violet 4BN : 15 percent	acid	698(NAC)
Glauber Salt : 69 percent		

The two (2) dyes (Alphazurine A and Wool Violet 4BN) are strong colours. For this Exfoliation test, Glauber salt could be omitted and less dye required to obtain the desired colour for the test. In that case, the following proportions are suggested:

	Class	Colour Index
Wool Violet 4BN : 48 percent	acid	698(NAC)
Alphazurine A : 52 percent	acid	714(NAC)

Table 38-1 - ISA classification table for raw silk of category I (18 denier and finer)

Major items / Grade	4A	3A	2A	A	B
Size Deviation (denier) 12 d. and below	0.80	0.95	1.10	1.35	above 1.35
13 d. - 15 d	0.90	1.05	1.25	1.50	above 1.50
16 d. - 18 d.	1.00	1.20	1.40	1.70	above 1.70
Evenness Variation I (count)	150	170	190	210	above 210
Evenness Variation II (count)	10	17	26	37	above 37
Cleanness (%)	97	95	93	88	below 88
Average Neatness (%)	94	92	90	87	below 87
Low Neatness (%)	90	87	83	77	below 77
Auxiliary / Class	(1)	(2)	(3)	(4)	(5)
Maximum Deviation (denier) 12 d. and below	2.2	2.6	3.0	3.6	above 3.6
13 d. - 15 d	2.4	2.8	3.3	4.1	above 4.1
16 d. - 18 d.	2.7	3.2	3.8	4.6	above 4.6
Evenness Variation III (count)	0	1	2	6	above 6
Auxiliary / Class	(1)	(2)		(3)	(4)
Winding (breaks) 12 d. and below	7	15		25	above 25
13 d. - 18 d.	5	12		21	above 21
Auxiliary / Class	(1)				(2)
Tenacity (grams)	3.7				below 3.7
Elongation (%)	18				below 18
Cohesion (strokes)	40				below 40

Table 38-2 - ISA classification table for raw silk of category II

Major items	Grade	4A	3A	2A	A	B
Size Deviation (denier)	19 d. - 22 d	1.15	1.35	1.60	1.95	above 1.95
	23 d. - 25 d.	1.30	1.50	1.80	2.20	above 2.20
	26 d. - 29 d.	1.40	1.65	1.95	2.35	above 2.35
	20 d. - 33 d	1.50	1.75	2.05	2.50	above 2.50
Evenness Variation I (count)		150	170	190	210	above 210
Evenness Variation II (count)		10	17	26	37	above 37
Cleanness (%)		97	95	93	88	below 88
Average Neatness (%)		94	92	90	87	below 87
Low Neatness (%)		90	87	83	77	below 77
Auxiliary	Class	(1)	(2)	(3)	(4)	(5)
Maximum Deviation (denier)	19 d. - 22 d.	3.1	3.6	4.3	5.3	above 5.3
	23 d. - 25 d.	3.5	4.1	4.9	5.9	above 5.9
	26 d. - 29 d.	3.8	4.5	5.3	6.3	above 6.3
	30 d. - 33 d.	4.0	4.7	5.5	6.8	above 6.8
Evenness Variation III (count)		0	1	2	6	above 6
Auxiliary	Class	(1)	(2)		(3)	(4)
Winding (breaks)		4	10		18	above 18
Auxiliary	Class	(1)				(2)
Tenacity (grams)		3.7				below 3.7
Elongation (%)		18				below 18
Cohesion (strokes)		60				below 60

Table 38-3 - ISA classification table for raw silk of category III
(34 denier and coarser)

Major items \ Grade		4A	3A	2A	A	B
Size Deviation (denier)	34 d. - 49 d	2.60	3.10	3.65	4.45	above 4.45
	50 d. - 69 d.	3.75	4.40	5.20	6.35	above 6.35
	70 d. and above	4.45	5.25	6.20	7.60	above 7.60
Maximum Deviation (denier)	34 d. - 49 d.	8.0	9.5	11.0	13.5	above 13.5
	50 d. - 69 d.	11.0	13.0	15.5	19.0	above 19.0
	70 d. and above	13.5	16.0	18.5	23.0	above 23.0
Evenness Variation I (count)		150	170	190	210	above 210
Evenness Variation II (count)		10	17	26	37	above 37
Cleanness (%)		97	95	93	88	below 88
Average Neatness (%)		94	92	90	87	below 87
Low Neatness (%)		90	87	83	77	below 77
Auxiliary \ Class		(1)	(2)	(3)	(4)	(5)
Evenness Variation III (count)		0	1	2	6	above 6
Auxiliary \ Class		(1)	(2)		(3)	(4)
Winding (breaks)	34 d. - 69 d.	1	6		13	above 13
	70 d. and above	0	4		10	above 10
Auxiliary \ Class		(1)				(2)
Tenacity (grams)		3.7				below 3.7
Elongation (%)		18				below 18

Table 39 - Chinese classification table for raw silk

Major items	Grade	6A	5A	4A	3A	2A	A	B	C	D	E	F
Size Deviation d (dtex)	12d (13.3dtex) below		0.75 (0.83)	0.85 (0.94)	1.00 (1.11)	1.15 (1.28)	1.30 (1.44)	1.50 (1.67)	1.75 (1.94)	2.00 (2.22)		
	13 - 15d (14.4 - 16.7 dtex)		0.80 (0.89)	0.90 (1.00)	1.05 (1.17)	1.20 (1.33)	1.35 (1.50)	1.55 (1.72)	1.80 (2.00)	2.05 (2.28)		
	16 - 18d (17.8 - 20.0 dtex)		0.90 (1.00)	1.05 (1.17)	1.20 (1.33)	1.40 (1.56)	1.60 (1.78)	1.80 (2.00)	2.10 (2.33)	2.40 (2.67)		
	19 - 22d (21.1 - 24.4 dtex)	0.90 (1.00)	1.05 (1.17)	1.20 (1.33)	1.35 (1.50)	1.55 (1.72)	1.80 (2.00)	2.05 (2.28)	2.35 (2.61)	2.70 (3.00)		
	23 - 25d (25.6 - 27.8 dtex)	1.05 (1.17)	1.20 (1.33)	1.35 (1.50)	1.55 (1.72)	1.80 (2.00)	2.05 (2.28)	2.35 (2.61)	2.70 (3.00)	3.10 (3.44)		
	26 - 29d (26.9 - 27.8 dtex)	1.15 (1.28)	1.30 (1.44)	1.50 (1.67)	1.75 (1.94)	2.00 (2.22)	2.30 (2.56)	2.65 (2.94)	3.05 (3.39)	3.50 (3.89)		
	30 - 33d (33.3 - 36.7 dtex)	1.25 (1.39)	1.45 (1.61)	1.65 (1.83)	1.90 (2.11)	2.20 (2.44)	2.50 (2.78)	2.85 (3.17)	3.25 (3.61)	3.75 (4.17)		
	34 - 49d (37.8 - 54.4 dtex)		2.20 (2.44)	2.60 (2.89)	3.10 (3.44)	3.65 (4.06)	4.30 (4.78)	5.05 (5.61)	5.95 (6.61)	7.00 (7.78)		
	50 - 69d (55.6 - 76.7 dtex)		3.15 (3.50	3.75 (4.17)	4.40 (4.89)	5.20 (5.78)	6.15 (6.83)	7.25 (8.06)	8.60 (9.56)	10.00 (11.11)		
	70d (77.8 dtex) above		3.75 (4.17)	4.45 (4.94)	5.25 (5.83)	6.20 (6.89)	7.30 (8.11)	8.65 (9.61)	10.00 (11.11)	12.00 (13.33)		
Evenness Variation II	18d (20.0 dtex) below		8	14	22	32	44	58	74	92		
	19 - 33d (21/1 - 36.7 dtex)	2	4	8	14	22	32	44	58	74		
	34d (37.8 dtex) above		2	4	8	14	22	32	44	58		
Cleanness (%)		98	97	96	95	93	90	87	84	81	78	75
Nearness (%)		95	94	92	90	88	86	84	82	80	77	73
Maximum Size Deviation d (dtex)	34 - 49d (37.8 - 54.4 dtex)		7.0 (7.8)	8.0 (8.9)	9.5 (10.6)	11.0 (12.2)	13.0 (14.4)	15.0 (16.7)	17.5 (19.4)	21.0 (23.3)		
	50 - 69d (55.6 - 76.7 dtex)		9.5 (10.6)	11.0 (12.2)	13.0 (14.4)	15.5 (17.2)	18.5 (20.6)	21.5 (23.9)	25.5 (28.3)	30.0 (33.3)		
	70d (77.8 dtex) above		11.0 (12.2)	13.0 (14.4)	15.5 (17.2)	18.5 (20.6)	21.5 (23.9)	25.5 (28.3)	30.0 (33.3)	36.0 (40.0)		

Table 39 (continued)

Auxiliary		(1)	(2)	(3)	(4)	(5)	(6)	(7)
	12d (13.3 dtex) below		2.1 (2.33)	2.1 (2.67)	2.8 (3.11)	3.4 (3.78)	4.6 (5.11)	4.6 above (5.11
	13 - 15d (14.4 - 16.7 dtex)		2.2 (2.44)	2.5 (2.78)	2.9 (3.22)	3.6 (4.00)	4.7 (5.22)	4.7 above (5.22)
Maximum	16 - 18d (17.8 - 20.9 dtex)		2.5 (2.78)	2.9 (3.22)	3.4 (3.78)	4.2 (4.67),	5.5 (6.11)	5.5 above (6.11)
Size deviation	19 - 22d (21.1 - 24.4 dtex)	2.5 (2.78)	2.9 (3.22)	3.4 (3.78)	3.8 (4.22)	4.7 (5.22)	6.2 (6.89)	6.2 above (6.89)
d (dtex)	23 - 25d (25.6 - 27.8 dtex)	2.9 (3.22)	3.4 (3.78)	3.8 (4.22)	4.3 (4.78)	5.4 (6.00)	7.1 (7.89)	7.1 above (7.89)
	26 - 29d (28.9 - 32.2 dtex)	3.2 (3.56)	3.6 (4.00)	4.2 (4.67)	4.9 (5.44)	6.6 (6.67)	8.9 (8.89)	8.0 above (8.89)
	30 - 33d (33.3 - 36.7 dtex)	3.5 (3.89)	4.1 (4.56)	4.6 (5.11)	5.6 (5.89)	6.6 (7.33)	8.5 (9.44)	8.5 above (9.44)

Auxiliary / Class	(1)	(2)	(3)	(4)	(5)	(6)	(7)
Evenness Variation I	100	120	140	160	180	220	220 above

Auxiliary / Class	(1)	(2)	(3)	(4)	(5)	(6)
Evenness Variation III	0	1	2	4	6	6 above

Auxiliary / Class		(1)	(2)	(3)	(4)	(5)	(6)
Winding (breaks)	12d (13.3 dtex) below		18	24	30	40	40 above
	13 - 18d (14.4 - 20.0 dtex)		14	20	26	36	36 above
	19 - 33d (21.1 - 36.7 dtex_		10	16	22	30	30 above
	34 - 69d (37.8 - 76.7 dtex)		4	8	12	16	16 above
	70d (77.8 dtex) above		2	4	8	12	12 above

Auxiliary / Class		(1)		(2)	(3)
Tenacity (grams)		3.70		3.60	3.60 below
Elongation (%)		19.0		18.0	18.0 below
Cohesion (strokes)	18d(20.0 dtex) below	60		50	50 below
	19 - 33d (21.1 - 36.7 dtex)	90	80	70	70 below

Table 40-1. Japanese classification table for raw silk of category I. (18 denier and finer)

Major	Grade	5A	4A	3A	2A	A	B	C	D
Size Deviation (denier)	12d and below	0.79 below	0.80 below	0.95 below	1.10 below	1.35 below	1.75 below	2.35 below	2.35 below
	13-15d	0.75 below	0.90 below	1.05 below	1.25 below	1.50 below	1.95 below	2.60 below	2.60 below
	16-18d	0.85 below	1.00 below	1.20 below	1.40 below	1.70 below	2.20 below	2.95 below	2.95 below
Evenness Variation II (count)	12d and below	17 below	26 below	37 below	50 below	65 below	82 below	101 below	101 below
	13-18d	10 below	17 below	26 below	37 below	50 below	65 below	82 below	82 below
Cleanness and Neatness (%)		98 above	97 above	95 above	93 above	88 above	79 above	66 above	66 below
Auxiliary	Class	(1)	(2)	(3)	(4)	(5)	(6)	(7)	(8)
Maximum Deviation (denier)	12d and below	1.9 below	2.2 below	2.6 below	3.0 below	3.6 below	4.7 below	6.3 below	6.3 above
	13-15d	2.0 below	2.4 below	2.8 below	3.3 below	4.1 below	5.3 below	7.0 below	7.0 above
	16-18d	2.3 below	2.7 below	3.2 below	3.8 below	4.6 below	5.9 below	8.0 below	8.0 above
Auxiliary	Class	(1)		(2)	(3)	(4)		(5)	(6)
Evenness Variation III		1 below		2 below	6 below	13 below		22 below	22 above
Auxiliary	Class	(1)		(2)		(3)		(4)	(5)
Winding (breaks)	12d and below	7 below		15 below		25 below		37 below	37 above
	13-18d	5 below		12 below		21 below		32 below	32 above
Auxiliary	Class	(1)							(2)
Tenacity (g)		3.7 above							3.7 below
Auxiliary	Class	(1)		(2)					(3)
Elongation (%)		19 above		18 above					18 below

Table 40-2 Japanese classification table for raw silk of category II. (19 to 33 denier)

Major items	Grade	5A	4A	3A	2A	A	B	C	D
Size Deviation (denier)	19-22d	1.00 below	1.15 below	1.35 below	1.60 below	1.95 below	2.50 below	3.35 below	3.35 above
	23-25d	1.10 below	1.30 below	1.50 below	1.80 below	2.20 below	2.80 below	3.75 below	3.75 above
	26-29d	1.15 below	1.35 below	1.60 below	1.90 below	2.30 below	2.95 below	3.95 below	3.95 above
	30-33d	1.30 below	1.50 below	1.75 below	2.05 below	2.50 below	3.20 below	4.30 below	4.30 above
Evenness Variation II (count)	19-25d	5 below	10 below	17 below	26 below	37 below	50 below	65 below	65 above
	26-33d	3 below	7 below	13 below	20 below	30 below	42 below	56 below	56 above
Cleanness and Neatness (%)		98 above	97 above	95 above	93 above	88 above	79 above	66 above	66 below
Auxiliary	Class	(1)	(2)	(3)	(4)	(5)	(6)	(7)	(8)
Maximum Deviation (denier)	19-22d	2.7 below	3.1 below	3.6 below	4.3 below	5.3 below	6.8 below	9.0 below	9.0 above
	23-25d	3.0 below	3.5 below	4.1 below	4.9 below	5.9 below	7.6 below	10.1 below	10.1 above
	26-29d	3.1 below	3.6 below	4.3 below	5.1 below	6.2 below	8.0 below	10.7 below	10.7 above
	30-22d	3.5 below	4.0 below	4.7 below	5.5 below	6.8 below	8.6 below	11.6 below	11.6 above
Auxiliary	Class	(1)		(2)	(3)	(4)		(5)	(6)
Evenness Variation III		0		1 below	2 below	6 below		13 below	13 above
Auxiliary	Class	(1)		(2)		(3)		(4)	(5)
Winding (breaks)	19-25d	4 below		10 below		18 below		29 below	29 above
	26-33d	3 below		9 below		17 below		27 below	27 above
Auxiliary	Class	(1)							(2)
Tenacity (g)		3.7 above							3.7 below
Auxiliary	Class	(1)	(2)	(3)					(4)
Elongation (%)		20 above	19 above	18 above					18 below

Table 40-3 Japanese classification table for raw silk of category III. (34 denier and coarser)

Major items	Grade	5A	4A	3A	2A	A	B	C	D
Size Deviation (denier)	34-49d	2.30 below	2.60 below	3.10 below	3.65 below	4.45 below	5.70 below	7.65 below	7.65 above
	50-69d	3.25 below	3.75 below	4.40 below	5.20 below	6.35 below	8.15 below	10.90 below	10.90 above
	70 above	3.90 below	4.45 below	5.25 below	6.20 below	7.60 below	9.75 below	13.05 below	13.05 above
Maximum Deviation (denier)	34-49d	7.0 below	8.0 below	9.5 below	11.0 below	13.5 below	17.0 below	23.0 below	23.0 above
	50-60d	10.0 below	11.0 below	13.0 below	15.5 below	19.0 below	24.5 below	32.5 below	32.5 above
	70 above	11.5 below	13.5 below	16.0 below	18.5 below	23.0 below	29.0 below	39.0 below	39.0 above
Evenness Variation II (count)		1 below	3 below	7 below	13 below	20 below	30 below	42 below	42 above
Cleanness and Neatness (%)		98 above	97 above	95 above	93 above	88 above	79 above	66 above	66 below
Auxiliary	Class	(1)				(2)	(3)	(4)	(5)
Evenness Variation III		0				1 below	2 below	6 below	6 above
Auxiliary	Class	(1)		(2)		(3)		(4)	(5)
Winding (breaks)	34-49d	1 below		6 below		13 below		22 below	22 above
	70 above	0		4 below		10 below		18 below	18 above
Auxiliary	Class	(1)							(2)
Tenacity (g)		3.7 above							3.7 below
Auxiliary	Class	(1)	(2)	(3)					(4)
Elongation (%)		20 above	19 above	18 above					18 below

Table 41-1 Indian classification table for Class I raw silk
[2.0 tex (or 18 denier) and finer]

Major items / Grade	6A	5A	4A	3A	2A	A	B	C	D
Size Deviation, (tex or denier)									
1.3 tex (or 12d)	0.089	0.094	0.106	0.117	0.128	0.130	0.150	0.167	above 0.167
(or 12d) and below	(0.80)	(0.85)	(0.95)	(1.05)	(1.15)	(1.25)	((1.35)	(1.50)	(above 1.50)
1.4 to 1.7 tex	0,100	0.106	0.117	0.128	0.139	0.150	0.167	0.189	above 0.189
(or 13 to 15d)	(0.90)	(0.95)	(1.05)	(1.15)	(1.25)	(1.35)	(1.50)	(1.70)	(above 1.70)
1.8 to 2.0 tex	0.117	0.128	0.139	0.150	0.161	0.178	0.194	0.217	above 0.217
(or 16 to 18d)	(1.05)	(1.15)	(1.25)	(1.35)	(1.45)	(1.60)	(1.75)	(1.95)	(above 1.95)
Evenness (%)	94	93	91	89	86	84	82	80	below 80
Low Evenness (%)	87	85	83	80	77	75	73	70	below 70
Cleanness (%)	96	95	94	93	92	90	88	85	below 85
Neatness (%)	95	94	93	92	90	88	86	84	below 84
Low Neatness (%)	92	90	88	86	83	79	75	70	below 70
Auxiliary / Grade	(I)	(II)	(III)	(IV)	(V)	(VI)	(VII)	(VIII)	(IX)
Maximum Deviation, (tex or denier)									
1.4 to 1.7 tex	0.23	0.26	0.28	0.30	0.33	0.37	0.40	0.44	above 0.44
(or 12d) and below	(2.1)	(2.3)	(2.5)	(2.7)	(3.0)	(3.3)	(3.6)	(4.0)	(above 4.0)
1.4 to 1.7 tex	0.27	0.28	0.30	0.33	0.37	0.40	0.44	0.50	above 0.50
(or 13 to 15d)	(2.4)	(2.5)	(2.7)	(3.0)	(3.3)	(3.6)	(4.0)	(4.5)	(above 4.5)
1,8 to 2.0 tex	0.31	0.33	0.37	0.40	0.43	0.47	0.51	0.58	above 0.58
(or 16 to 18d)	(2.8)	(3.0)	(3.3)	(3.6)	(3.9)	(4.2)	(4.6)	(5.2)	(above 5.2)
Auxiliary / Grade		(I)		(ii)			(III)		(IV)
Winding (breaks)									
1.3 tex (or 12d) and below		10		15			23		above 23
1.4 to 2.0 tex (or 13 to 18d)		7		12			20		above 20
Auxiliary / Grade		(I)				(II)			(III)
Tenacity g/tex (or g/denier)		33 (3.7)				32 (3.6)			below 32 (below 3.6)
Elongation (%)		19				18			(below 18)
Cohesion (strokes) 1.4 to 2.0 tex (or 13 to 18d)		40				35			(below 35)
Auxiliary / Grade			(I)					(II)	
Cohesion (strokes) 1.3 tex (or 12d) and below			30					below 30	

Table 41-2 Indian classification table for Class I raw silk (2.1 to 3.7 tex or 19 to 33 denier)

Major items / Grade	6A	5A	4A	3A	2A	A	B	C	D	E
Size Deviation tex (or denier)										
2.1 to 2.4 tex	0.128	0.139	0.150	0.167	0.183	0.200	0.217	0.239	0.267	above 0,.267
(or 19 to 22d)	(1.15)	(1.25)	(1.35)	(1.50)	(1.65)	(1.80)	(1.95)	(2.15)	(2.40)	(above 2.40)
2.6 to 3.0 text	0.156	0.167	0.183	0.200	0.217	0.233	0.256	0.278	0.300	above 0.300
(or 23 to 27d)	(1.40	(1.50)	(1.65	(1.80)	(1.95)	(2.10)	(2.30)	(2.50)	(2.70)	(above 2.70)
3.1 to 3.7 tex	0.178	0.194	0.211	0.233	0.256	0.278	0.300	0.328	0.356	above 0.356
(or 28 to 33d)	(1.60	(1.75)	(1.90)	(2.10)	(2.10)	(2.50)	(2.70)	(2.95)	(3.20)	(above 3.20)
Evenness (%)	94	93	91	89	86	84	82	80	77	below 77
Low Evenness (%)	87	85	83	80	77	75	73	70	66	below 66
Cleanness (%)	96	95	94	93	92	90	88	85	81	below 82
Neatness (%)	95	94	93	92	90	88	86	84	82	below 82
Low Neatness (%)	92	90	88	86	83	79	75	70	64	below 64
Auxiliary / Grade	**(I)**	**(II)**	**(III)**	**(IV)**	**(V)**	**(VI)**	**(VII)**	**(VIII)**	**(IX)**	**(X)**
Maximum Deviation, tex (or denier)										
2.1 to 2.4 tex	0.33	0.37	0.40	0.44	0.49	0.53	0.58	0.63	0.71	above 0.71
(or 19 to 22d)	(3.0)	(3.3)	(3.6)	(4.0)	(4.4)	(4.6)	(5.2)	(5.7)	(6.4)	(above 6.4)
2.6 to 3.0 tex	0.42	0.46	0.49	0.53	0.58	0.63	0.69	0.76	0.82	above 0.82
(or 23 to 27d)	(3.8)	(4.1)	(4.4)	(4.8)	(5.2)	(5.7)	(6.2)	(6.8)	(7.4)	(above 7.4)
3.1 to 3.7 tex	0.48)	0.52	0.58	0.63	0.69	0.74	0.81	0.89	0.97	above 0.97
(or 28 to 33d)	(4.3)	(4.7)	(5.2)	(5.7)	(6.2)	(6.7)	(7.3)	(8.0)	(8.7)	(above 8.7)

Auxiliary / Grade	(I)	(II)	(III)	
Winding (breaks) above 20	6	10	15	20

Auxiliary / Grade	(I)	(II)	(III)
Tenacity g/tex (or g/denier)	33 (3.7)	32 (3.6)	below 32 (below 3.6)
Elongation (%)	19	18	below 18
Cohesion (strokes)	60	50	below 50

Table 41-3 Indian classification table for Class I raw silk (3.8 tex or 34 denier and coarser)

Major items \ Grade	4A	3A	2A	A	B	C	D	E
Size Deviation, tex (or denier)								
3.8 to 5.4 tex	0.344	0.389	0.433	0.489	0.556	0.656	0.778	above 0.778
(or 34 to 49d)	(3.10)	(3.50)	(3.90)	(4.40)	(5.00)	(5.90)	(7.00)	(above 7.00)
5.6 to 7.7 tex	0.456	0.511	0.578	0.644	0.744	0.878	1.033	above 1.033
(or 50 to 69d)	(4.10)	(4.60)	(5.20)	(5.80)	(6.70)	(7.90)	(9.30)	(above 9.30)
7.8 tex	0.567	0.633	0.700	0.789	0.911	1.078	1.267	above 1.267
(or 70d) and above	(5.10)	(5.70)	(6.30)	(7.10)	(8.20)	(9.70)	(11.40)	(above 11.40)
Maximum Deviation, tex or denier								
3.8 to 5.4 tex	1.00	1.11	1.22	1.44	1.67	2.00	2.33	above 2.33
(or 34 to 49d)	(9.0)	(10.0)	(11.0)	(13.0)	(15.0)	(18.0)	(21.0)	(above 21.0)
5.6 to 7.7 tex	1.33	1.56	1.78	2.00	2.33	2.67	3.11	above 3.11
(or 50 to 69d)	(12.0)	(14.0)	(16.0	(18.0)	(21.0)	(24.0)	(28.0)	(above 28.0)
7.8 tex	1.67	1.89	2.11	2.44	2.78	3.22	3.78	above 3.78
(or 70d) and above	(15.0)	(17.0)	(19.0)	(22.0)	(25.0)	(29.0)	(34.0)	(above 34.0)
Evenness (%)	91	89	86	84	82	80	77	below 77
Low Evenness (%)	83	80	77	775	73	70	66	below 66
Cleanness (%)	94	92	90	87	83	79	75	below 75
Neatness (%)	93	91	89	87	84	81	78	below 78
Low Neatness (%)	87	85	82	78	74	68	62	below 62

Auxiliary \ Grade	(I)	(II)	(III)	(IV)	(V)
Winding (breaks)					
3.8 to 7.7 tex (or 34 to 69d)	3	6	10	15	above 15
7.8 tex (or 70d) and above	2	4	6	10	above 10

Auxiliary \ Grade	(I)	(II)	(III)
Tenacity b/tex	33	32	below 32
(or g/denier)	(3.7)	(3.6)	(below 3.6)
Elongation (%)	19	18	below 18

References

Auyzawa, C. *et al.* 1972. *Handbook of silkworm rearing.* Japan, Fuji Publishing Co. Ltd.

Byong Hee Choe. 1973. *The outline of natural silk processing.* Republic of Korea, Korea Silk Engineering Society.

Byong Hee Choe. 1978. *Sericulture technology.* Republic of Korea, Seoul National University Press.

Byong Ho Kim. 1983. *Filature water engineering.* Republic of Korea, Seoul Publishing Co.

Choe, B.H., *et al.* 1969. *Studies of Eri-silk culturing in Korea.* Republic of Korea, *Sericulture Journal of Korea*, Vol. 9.

Choe, B.H., Song, K.E., Kim, H.N. and Lee, Y.W. 1971. *The experiment for the technical balance work on the automatic silk reeling machine.* Republic of Korea, *Sericulture Journal of Korea*, Vol. 13(2).

Economic and Social Commission for Asia and the Pacific. 1992. *Proceedings of the Expert Group Meeting on Post-cocoon operations.* Manila, the Philippines, 26-29 October 1992. United Nations.

Economic Commission for Asia and the Pacific. 1994. *Silk in Asia.* United Nations, New York, USA.

FAO. 1972. *Manual on sericulture, Vol. 3. Silk reeling.* Agricultural Services Bulletin No. 72/3.

Kim, G.K., Lee, I.J. and Choe, H. 1965. *Effect of reelability of aid reagents on the dupion silk reeling.* Republic of Korea, Annual Research Report of the Korean Sericulture Experiment Station.

Kimura, S. *et al.* 1964. *Lectures on silk reeling technology.* Japan. Cocoon Testing Committee, Japan Sericultural Association.

Kirimura N. 1957. *Recent analyses of silk protein. "Structures of cocoon silk filament*

Lee, .W. 1984. *Cocoon technology and raw silk reeling and testing consultancy report.* UNDP/FAO Project SRL/81/015.

National Sericultural Inspection Office. 1994. *Comparisons in the methods of raw silk testing among ISA, Japan, China and Korea.* Seoul, Republic of Korea.

National Sericulture and Entomology Research Institute. 1992. *Principles and practices in sericulture.* Republic of Korea, SERI/RDA.

Song, K.E. 1967. *Study on Tussah silk reeling method (V).* Research report of the Rural Development Administration (Sericulture), Vol. 10.

Sone, K.E. and Choe, S.R. 1970. *Effects of various silk reeling conditions on the occurrence of defects onto raw silk.* Research report of the Rural Development Administration (Sericulture), Vol. 13.

Song, K.E. and Kim, Y.H. 1974. *A study on the improvement of fresh cocoon transportation.* Research report of the Rural Development Administration (Sericulture), Vol. 16.

Song, K.E. and Lee, Y.W. 1973a. *Modern silk reeling technology.* Republic of Korea, Sericulture Experiment Station.

Song, K.E. and Lee, Y.W. 1973b. *Study on the adjustment of filature water conditions.* Research report of the Rural Development Administration (Sericulture), Vol. 16.

Song, K.E. and Lee, Y.W. 1973c. *Study on the improvement of filature water conditions.* Research report of the Rural Development Administration (Sericulture), Vol. 15.

Song, K.E. and Lee, Y.W. 1975a. *Studies on the modification of natural water conditions for cocoon cooking operations in silk reeling process.* Research report of the Rural Development Administration (Sericulture), Vol. 17.

Song, K.E. *et al.* 1975b. *Studies on the improvement of floss silk and hand-spun silk processing.* Research report of the Rural Development Administration (Sericulture and Veterinary), Vol. 17.

Song, K.E., Lee, Y.W. and Choe, B.H. 1973. *Cocoon cooking conditions for the technical balance work on automatic silk reeling machine.* Research report of the Rural Development Administration (Sericulture), Vol. 15.

APPENDIX

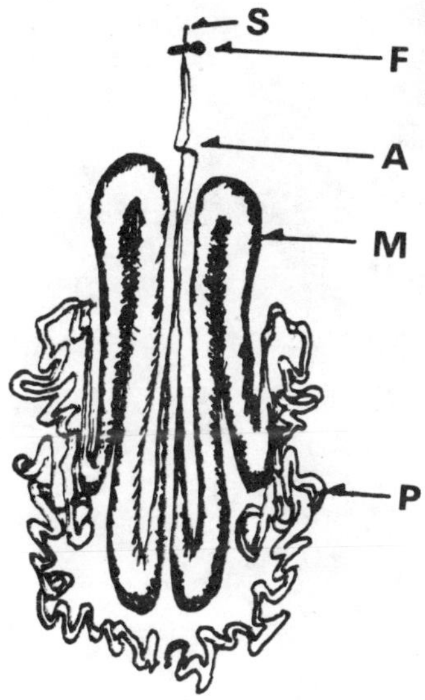

Figure 1. Silk glands of the larva of *Bombyx mori*
S: spinneret F: Filipi's gland A: anterior section
M: middle section P: posterior section

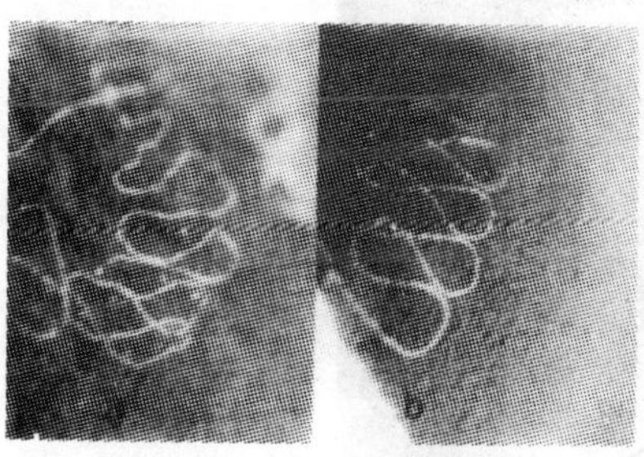

Figure 2. Cocoon spinning curvature
a: outer layer b: inner layer

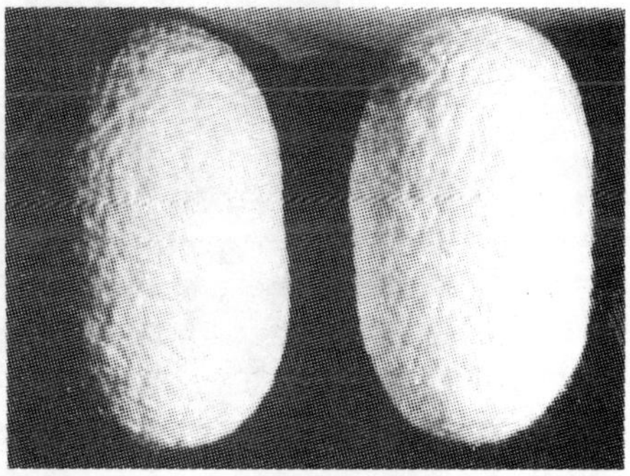

Figure 3. Bivoltine F_1 hybrid cocoons
(Japanese and Chinese)

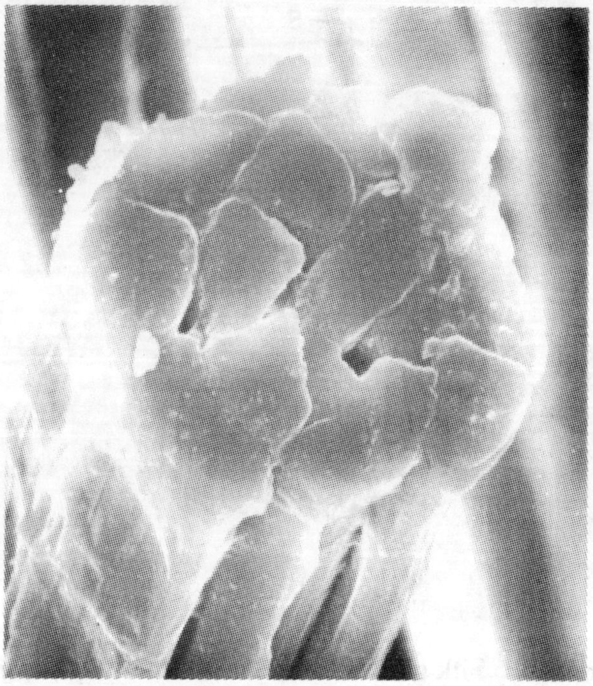

Figure 4. Cross section of raw silk thread (1,300 times by SEM)

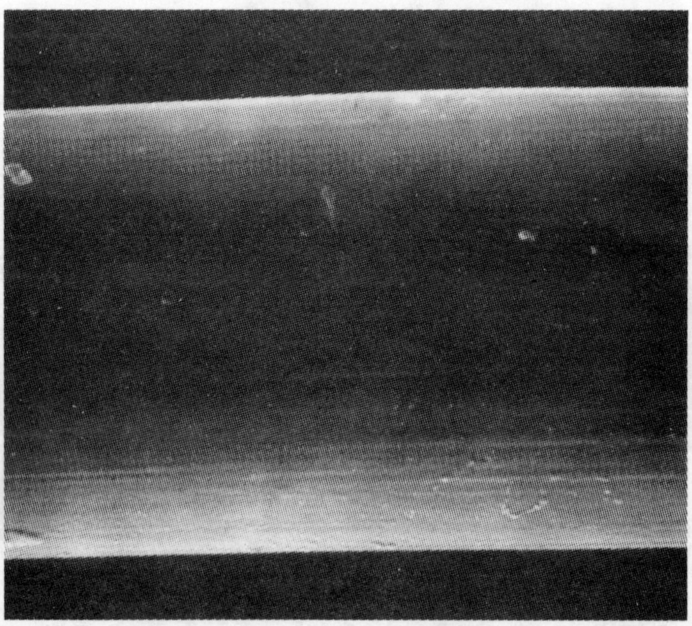

Figure 5. A brin of degummed cocoon bave (5,000 times by SEM)

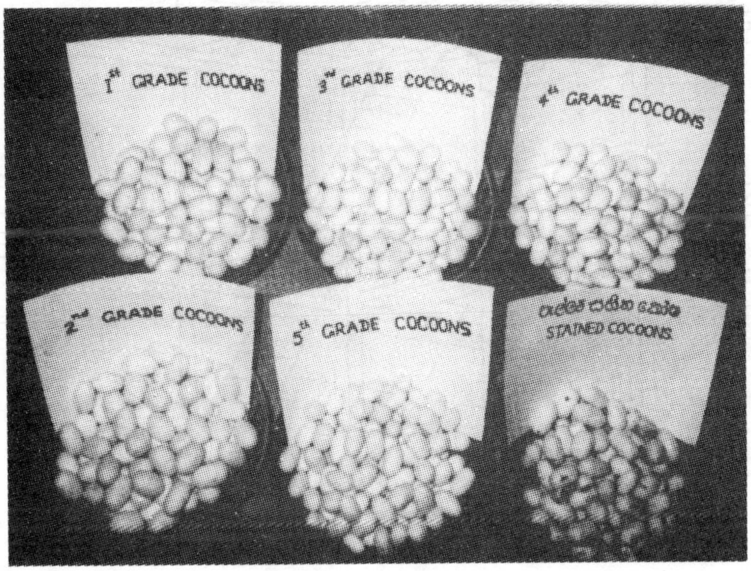

Figure 6. Cocoons grades by visual inspection

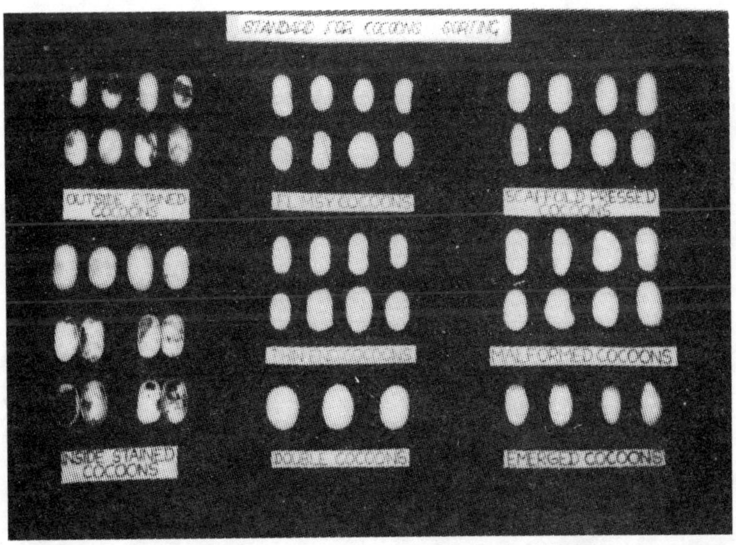

Figure 7. Standards of assorted cocoons

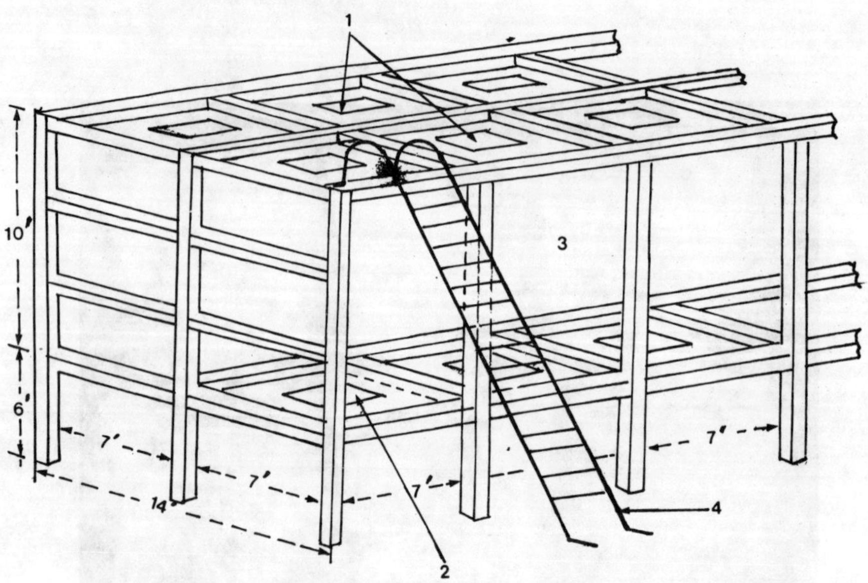

Figure 8. Dimension of "honey comb Tub" type cocoon storage
1) inlet 2) outlet 3) plywood double wall 4) aluminium ladder

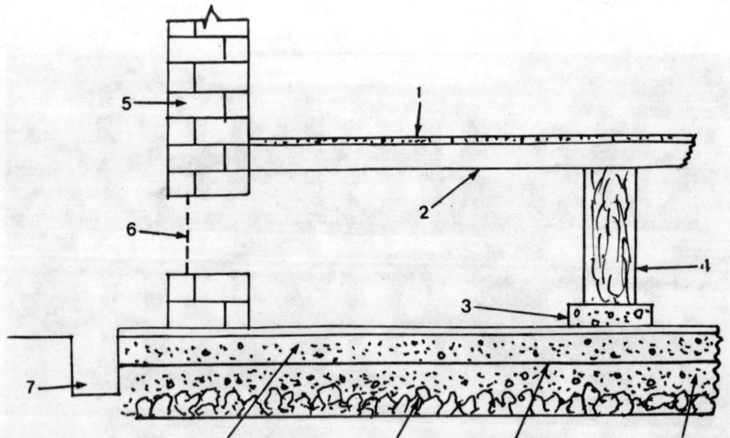

Figure 9. Cross section of cocoon storage
1) planks, 2) wooden beam 3) base 4) post 5) brick wall 6) wire net for ventilation 7) drain 8) polyethylene 9) primary concrete 10) secondary concrete 11) gravel

Figure 10. Cabinet type cocoon dryer (shelf-carrier type)

Figure 11. Hot air circulating type cocoon dryer (hot air blower)

Figure 12. Cocoon deflossing machine

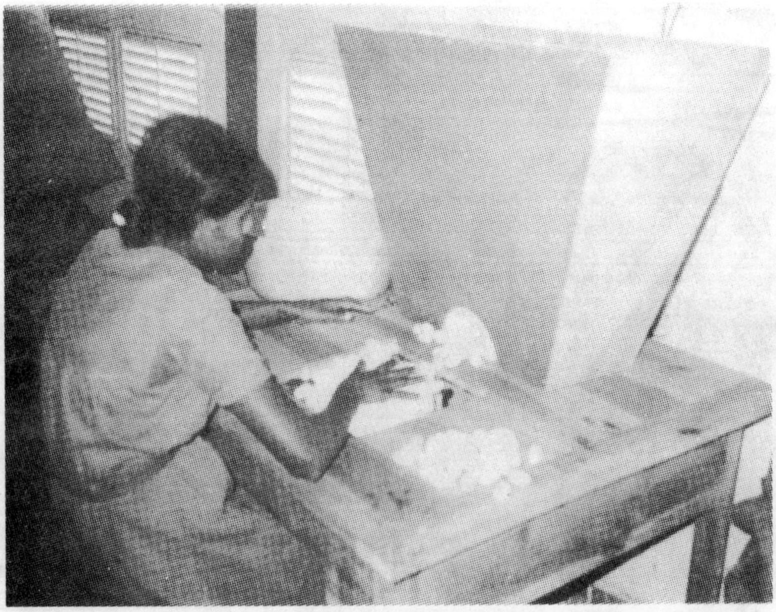

Figure 13. Cocoon assorting table

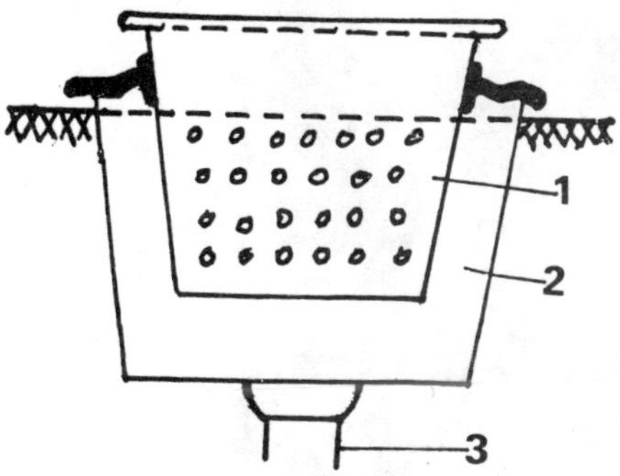

Figure 14. Pan cooking basin
1) inner perforated vessel
2) outer vessel
3) waste water drain

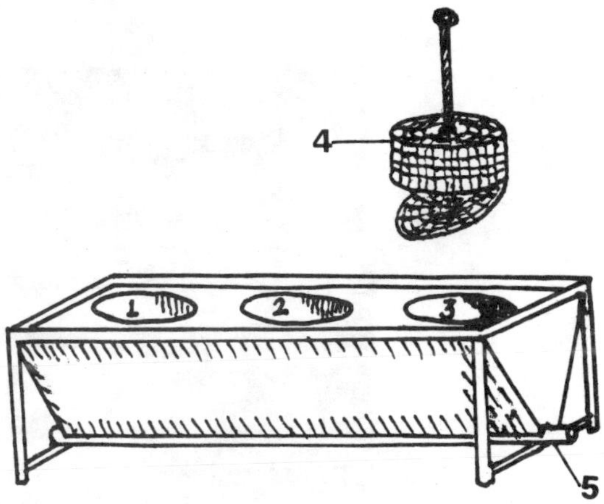

Figure 15. Three pan type cooking appliance
1) first bath
2) second bath
3) third bath
4) cocoon wire cage
5) drain pipe

Figure 16. Cocoon cooking machine (basket circulating type)

Figure 17. Treadle type reeling machine

Figure 18. Sitting type reeling machine (motor driven)

Figure 19. Multi-ends type reeling machine

Figure 20. Automatic reeling machine for cocoon testing (3 ends per basin)

Figure 21. Picking-ends part of automatic silk reeling machine

Figure 22. Automatic silk reeling machine

Figure 23. Vacuum pre-treatment for re-reel.

Figure 24. Raw silk re-reeling machine

Figure 25. Lacing of silk skeins

Figure 26. Booking appliance of silk skeins

Figure 27. Water softening equipment (for cooking machine)

Figure 28. Silk by-product treating machine

Figure 29. Floss silk prepared from double cocoons

Figure 30. Hand-spun silk reeling machine (from floss silk)

Figure 31. Automatic degumming machine of by-products for spun silk

Figure 32. Settling machine for spun silk

Figure 33. Conditioning oven of raw silk

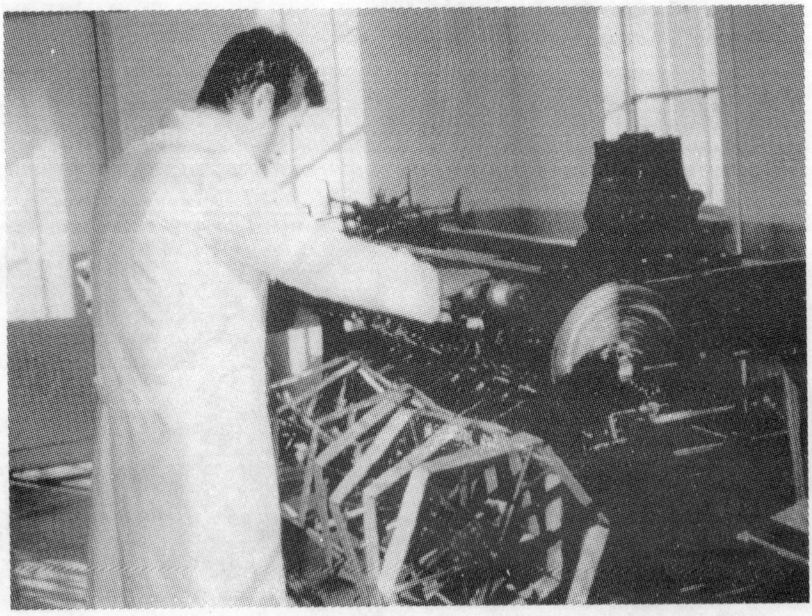

Figure 34. Winding tester of silk skein

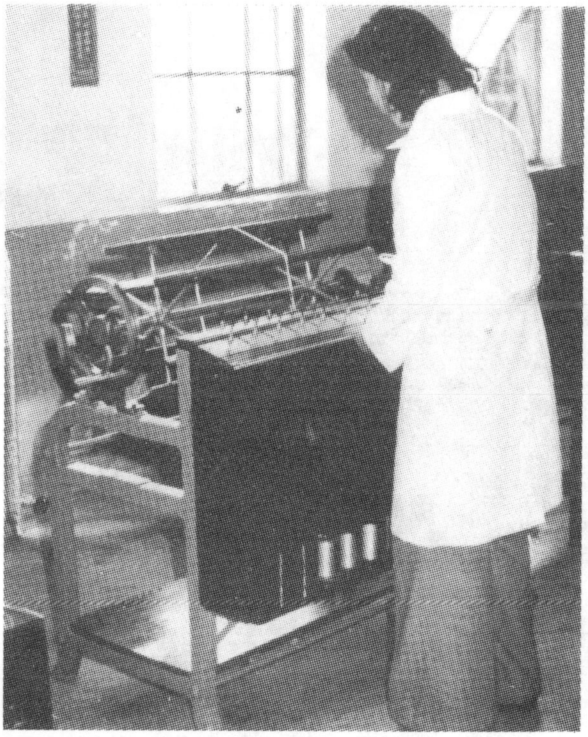

Figure 35. Small skein sampler for size deviation test

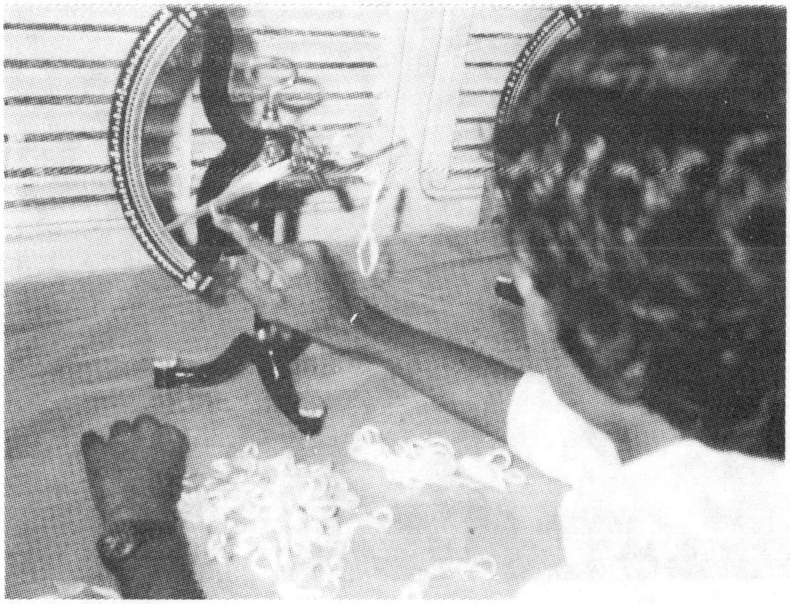

Figure 36. Denier balance

Figure 37. Seriplane winder

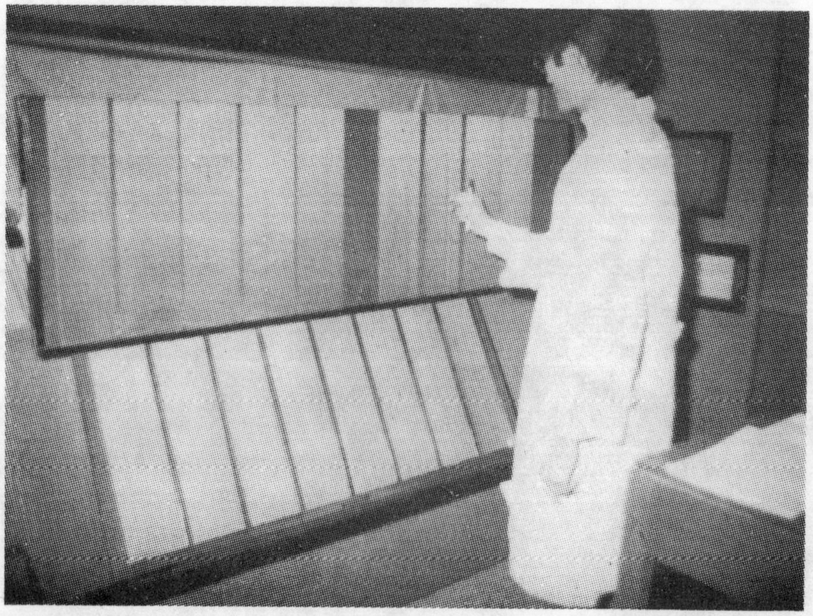

Figure 38. Seriplane inspection for cleanness and neatness

Figure 39. Duplan type cohesion tester

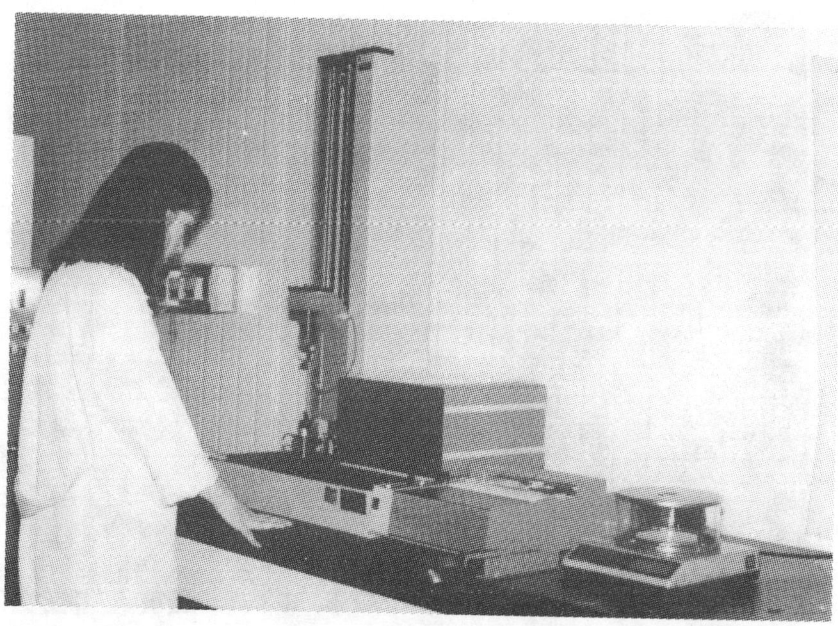

Figure 40. Tenacity and elongation tester

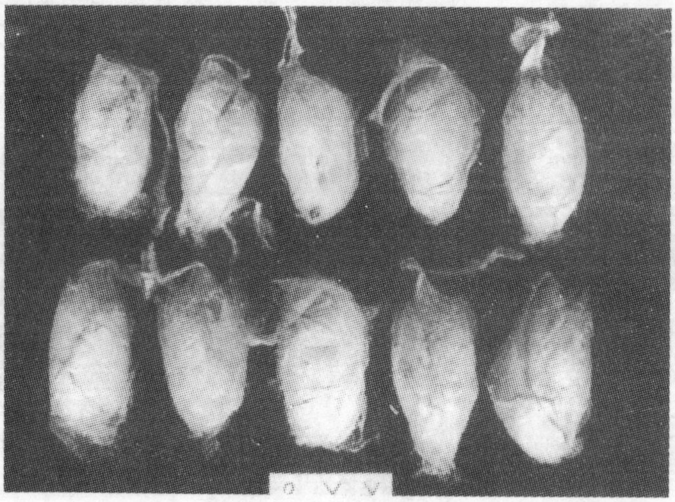

Figure 41. Tasar cocoons (*Antheraea pernyi*)

Figure 42. Wild cocoons (*Antheraea yamamai*)